A HISTORY OF
INDUSTRIAL
DESIGN

A HISTORY OF INDUSTRIAL DESIGN

Edward Lucie-Smith

PHAIDON ~ OXFORD

Phaidon Press Limited,
Littlegate House, St Ebbe's Street, Oxford OX1 1SQ

First published 1983
© 1983 by Phaidon Press Limited

British Library Cataloguing in Publication Data

Lucie-Smith, Edward
    A history of industrial design.
    1. Design, Industrial — History
    I. Title
    745.2'09   TS171

ISBN 0—7148—2281—7

Designed by Sarah Tyzack

Filmset by Keyspools Limited, Golborne, England

Printed in Great Britain
by William Collins Sons & Co. Limited

Frontispiece. Part of the superstructure of the P & O liner
Oronsay, 1951; cf. fig. 274.

The publishers would like to thank the following owners and copyright holders
for permission to reproduce photographs:
AEG Telefunken Firmenarchiv 56, 170, 171; Alexandria Press, London 27, 28;
Aram Designs 172; Architectural Press 182; Art and Artists 298; Arzberg,
Germany 176; Michael Auer 354; Bakelite Museum Society 379; Colin Banks
435–7; Bauhaus-Archiv 174; Bedford, Lemere & Co. Ltd. 139; Maria de Beyrie
186 (photo Christian Radux), 187; Hedrich Blessing Studio, Chicago 270;
Boothroyd-Stewart 380; Braun AG 384, 387 (photo S.R. Gnamm); British Airways
262; Courtesy of the Trustees of the British Museum 7, 8, 10, 11, 44; British
Architectural Library/RIBA 149; British Railways Board 243, 244, 247, 271;
British Telecom 107, 363–5, 368, 372, 373; Margaret Burke-White/Time Life Inc.
1936/Colorific 258; James Cahes 339; Canon (UK) Ltd. 356; The Caravan
Club/Bristol Industrial Museum 264; Cassina 143, 159, 306; Cheney and Cheney,
Art and the Machine (Whittlesey House, McGraw Hill Co., 1936) 189; Chrysler
Corporation 210; Citroën 208; Coca-Cola Export Corporation 75; Compute-
Antique 13; Hugh Conway 202, 216; A.C. Cooper/collection George Ingham 31;
Cowderoy & Moss Ltd. 427; Gillian Darley 35, 76, 77, 148; Design Council 22, 65,
66, 68, 73, 86, 101, 106, 111, 114, 131, 133, 135, 136, 137, 140, 147, 173, 178, 192,
199, 241, 263, 278, 287, 295 (collection Hazel Conway), 301, 303, 305, 321, 327,
336, 338, 340, 343, 367 (photo Richard Davies), 371, 378, 388, 389, 393, 397,
405–7, 411, 413, 423; Design Council Archive 3, 4, 20, 49, 51, 53, 55, 59, 60, 67, 71
(Martin Coombs), 83–4 (Alexandre Georges), 85 (collection Sue and David
Gentleman), 92, 98, 100, 107 (British Telecom), 122, 129, 175 (Franz Lazi, Jr.), 188,
191, 203, 206, 239, 242, 245, 246, 248, 267, 269, 274, 284 (Bruno Stefani), 285, 286,
293, 296, 297, 300 (Black Star), 302 (R. H. de Burgh-Galwey), 311, 312, 314
(Newton Chambers & Co.), 315 (British Gas Co.), 316 (Belling), 319 (General
Electric Co., Ltd.), 320 (Westinghouse Electrical Co.), 322 (Good Housekeeping),
328 (E.K. Cole Ltd.), 330, 331 (Goblin), 334, 341, 351, 369, 370, 374–6, 381, 382,
392, 395, 396, 398 (Manor Studios), 401, 424, 425; Devon Conversions Ltd. 265;
Disabilities Study Unit 432; Mary Evans Picture Library 86, 87, 104, 200, 443;
Ferranti Archives 329; Fiat (UK) Ltd 229; Fiat, Centro Storico 225, 226; Fischer
Fine Art Ltd. 138, 144, 145; Form International 180, 181; Robin Garton 193, 196;
Glass Manufacturers' Federation 99, 103; S.R. Gnamm 80, 109, 165; Richard
Hamilton 337; Hasselblad (GB) 355; De Havilland 261; Heal & Son 146; Hendon
and District Archaeological Society 150; Bengt Höglund 439; G. Holme,
Industrial Design and the Future (1934) 390, 391; Dennis Hooker, Clareville
Studios Ltd. 69; Hoover Ltd. 332, 333, 335; IBM Archives 428; Illustrated
London News 47; Image Photography 29–31, 54, 105 and 110 (collection S. Katz),
352; ItalDesign 224, 227, 228, 230–2, 234, 235–7; Johnson Wax 185; Sylvia Katz
45; John Knight 141; Kodak Ltd. 353, 358; Lucinda Lambton 70, 72, 74, 409, 410;
Laurent Sully Jaulmes 96; J. & L. Lobmeyr, Vienna 83; London Transport 240;
John Makepeace 279; Eric S. de Maré 281, 282, 377; Marshmallow (photo H.
Grosso) 362; Minolta UK 360, 361; Minton Archives 132; Morgan-Grampian 1,
78, 125–7; Museum of London 16; National Maritime Museum 276; National
Monuments Record 197; National Motor Museum, Beaulieu 89, 128, 204, 205,
209, 212 (R.R. Stuart Collection), 213 (photo B. Butcher), 214, 218, 219, 220–2,
366; Netherlands Railways 249; The Observer 307; Odhams Press 57; Olympus
Optical Co. (UK) 359; Oscar Woollens 299, 308, 309, 310; P & O 272, 273, 275;
Penhaligon's 421–3; Roger Pennington 215; Pullman-Standard 268; R.A.F.
Museum 254–7, 259, 260; Rotring 404; Royal Academy of Art 142; Royal
Aeronautical Society 238, 253; Sainsbury & Co. Ltd. 102, 408, 415–20, 426, 429,
431; Crown Copyright, Science Museum 25, 26, 32, 33, 41–3, 99, 108, 112, 113,
115–18, 123, 124, 250–2, 266, 317, 318, 344, 345, 400; Sony (UK) Ltd. 357, 383,
385, 386, 403; Sotheby's 95, 347, 348; Föreningen Svensk Form 17, 294, 438, 441;
Thorn Domestic Appliances 323–6; Tothill Press Ltd. 283; Victoria and Albert
Museum 91; Louis Vuitton 97; Wedgwood, Ltd. 48–52; D. and F. Wellby Ltd. 19;
World Health Organization 442.

Contents

For John George

ACKNOWLEDGEMENTS

Particular thanks are due to Roger Sears and Bernard Dod of
Phaidon Press—the former for scrutinizing my original
synopsis and making many helpful suggestions, the latter for
editing the manuscript so efficiently, and to Sylvia Katz for a
magnificent job of picture research and many suggestions. Also
to Walter Collins of Oscar Woollens, Dr Felice Cornacchia, Mike
Thorold-Palmer and Richard Vitali of Fiat; Dr Bruno Molineri of
ItalDesign; and Dr John George of *Art & Artists*.

1 American locomotive in general use *c.* 1857. Engraving from *The
Engineer*.

Introduction

Industrial design suffers from the fact that it has become one of the alternative religions of our century. Essentially, the definition of the meaning of the phrase ought to be a very simple matter—it is the business of determining the form of objects which are to be made by machines, rather than produced by hand. But how immense that range of objects now is, and what a multitude of different categories they occupy! Industrial design can concern itself with everything from a teacup to a jet aeroplane. Yet it is not a matter of diversity alone—there is also our feeling that the machine production of a whole series of objects which are not merely similar but identical puts the designer of those objects into a very different position from the person who designs objects which are produced by hand. The former is divorced from the actual business of making, while the latter probably (but not absolutely inevitably) remains very close to it. The industrial designer therefore stands aside from the physicality of the manufacturing process, yet is responsible for analysing and trying to make sense of it. He is responsible for what F. H. K. Henrion, President of the Society of Industrial Artists, called 'an ordering process, creating at its best an inspired, new and unique order from a state of chaos'.

But this is not all. Industrial design is not a neutral

occupation. It remains coloured with the moralism of a race of Victorian prophets and pioneers—the chief among them was John Ruskin—who reacted against what they saw as the intolerable waste and squalor of the Industrial Revolution. This moralism has since been carried over into situations where it is not always appropriate. The twentieth-century industrial designer is seen as the custodian of public taste, a person responsible for guiding the recalcitrant mass towards enlightenment. Herbert Read, one of the great propagandists for industrial design in the first half of this century, spoke of 'the conflict between ideal form and popular taste' as if this was something inevitable. It is easier to understand what industrial design really is, and how the concept has developed historically, if one makes a resolution to ignore the more hectoring kinds of design propaganda.

In this connection—the actual history of design and the emergence of industrial design as a recognized profession—it is worth recalling that the word 'industry' was itself quite slow to acquire the meaning with which we endow it today when we employ it in the context presupposed by this book. In French, for instance, the equivalent word 'industrie' appears no earlier than the eighteenth century, while the adjective 'industrielle' is first recorded in 1770. The paradox is that many centuries before this the idea that beauty in everyday objects was somehow linked to efficiency and appropriateness for use had already occurred to intelligent men. In Xenophon's *Memorabilia* Socrates is quoted as saying, in reply to Aristippus: 'Is a dung-basket beautiful then?—Of course, and a golden shield is ugly, if the one is well made for its special work and the other badly.'

People were also acquainted with the notion that use must often be allowed to dictate form. Francis Bacon wrote: 'Houses are built to live in, and not to look on; therefore let use be preferred before uniformity, except when both may be had.' Bacon's statement already seems to presuppose the kind of thought process outlined by Herbert Read in his essay, 'The Origins of Form in Art', which seems so typical of our century thanks to the emphasis it puts on functionalism. Read distinguishes three stages in the development of objects of utility: 'Namely (1) discovery of functional form, (2) refinement of the functional form to its maximum efficiency, and (3) refinement of the functional form in the direction of free or symbolic form.'

Long before the profession of industrial designer was invented, there were people who carried out the designer's function. Basically, they can be divided into two groups—the artisans and the architects. Artisan design evolved from direct work with tools and materials, and even, in the early stages of the Industrial Revolution, from direct work with ma-

2 Table-glasses by Philip Webb, made by James Powell & Sons at Whitefriars. Late 19th century. London, Victoria and Albert Museum.

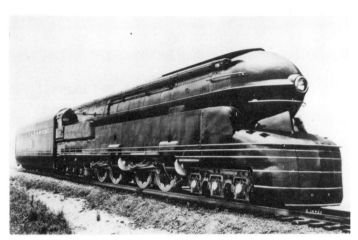

3 Pennsylvania Railroad S–1 Type locomotive, designed by Raymond Loewy.

chines and an intelligent exploration of their possibilities. Industry unconsciously evolved its own aesthetic, and this aesthetic was from the beginning one which intelligent designers, whatever their background, tried to obey. Philip Webb, the architect who built the seminal Red House for William Morris, and who later made designs for that fountainhead of the English Arts and Crafts Movement, the firm of Morris & Co., once said: 'I never begin to be satisfied until my work looks commonplace.' The table-glass which he designed for manufacture by James Powell & Sons at Whitefriars (fig. 2) proves that he meant what he said. The idea was thus planted very early that the successful industrially produced object eschewed not only ornament but anything which might make it seem conspicuous.

Yet it is also deeply significant that architecture was Webb's basic profession. The notion of a responsible designer who is separated from the craft process, but who still has the right to·instruct the craftsman, and tell him what to do, was established by the great architects of the eighteenth century, and particularly by Robert Adam, who designed the fittings and contents of some of his great country houses almost to the last detail. The furniture which Thomas Chippendale produced to Adam's designs was very different from the furniture he designed himself and published in his book *The Director*. The architect was a professional man, and it is from him that the industrial designer of our own day inherits his claim to professional status. Indeed, the two professions are still very often combined.

Industry, however, did not simply establish itself and then become a constant, a stable background against which the designer must work. It constantly threw up new problems. What happens, for example, when a machine is used to produce, not just some simple object, but another machine? Here, new-born, is a mechanism which may seem to dictate a surface which is visually complex, to reflect the complexity of a multitude of parts. Is the designer obliged to follow faithfully whatever lies beneath the casing? If he does, this may result in a form which is economic from the material point of view, but uneconomic visually because it requires a much greater effort of perception. If one looks at the career of a pioneer American designer, Raymond Loewy, one finds that much of what he did was an effort to solve this kind of problem. Often his proposals met with resistance from the engineers with whom he had to deal. Thus, when Loewy arrived to redesign the locomotives of the Pennsylvania Railroad, there was already an established notion in railway workshops about what a locomotive ought to look like—a tradition which stretched well back into the nineteenth century (fig. 1). To engineers, if not to the travelling public, a locomotive of traditional form seemed preferable to the streamlined design Loewy came up with (fig. 3), because for them it was far more expressive of the true nature of steam power.

Loewy was not himself a trained engineer, but someone who took over after the engineers had done their best or worst. This is a very common situation where industrial designers are concerned, and it calls into question the assertion made by one authority on the subject—that good design is 'the outward expression of the engineer's confidence in his work'. In fact, whether it is the creation of trained engineers or not, industrial design is quite often palliative, not radical. It is a technique that may be used to conceal faults, such as distortions in die-castings, introduced by the inaccuracy of machines, rather than to show off their accuracy. In these circumstances the industrial designer's job is to see that these inevitable faults do not spoil the finished result—for example by introducing a moulding to disguise an imperfect fit. In any case, the designer's task is often to establish limits rather than to conduct a search for perfection. He tries to trace the frontiers within which a range of acceptable solutions can be found. These boundaries are usually drawn for him by questions of cost as well as by those of structural strength and mechanical efficiency.

What else does the designer have to do in the real world? He must create objects which not only work as intended, but which clearly indicate what their function is—things which speak a visual language which anyone who is likely to use them will understand. This in turn means that the industrial designer has to deal with the way in which things are perceived, as well as the way in which they objectively exist. He must take into account both psychology and sociology.

There is a constant dialogue between mechanical invention on the one hand, and the way in which

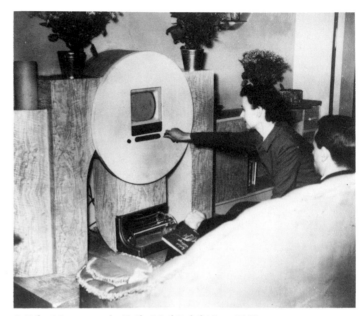

4 Television set at the Daily Mail Exhibition, 1949.

5 The Sinclair Microvision pocket television, 1981.

invention alters habits, manners and common assumptions on the other. Television supplies an excellent example. The designer of an early television set saw it as something which quite literally took the place of the traditional hearth—a focal point for the entire family (fig. 4). The designer of the Sinclair Microvision (fig. 5), which first became available in 1982, sees television in terms which are very different—his set is a sophisticated toy, and seems to presuppose a solitary, not a shared experience. The impetus for the change is partly sociological, but also partly technological. The Microvision exploits a technology—microchips, miniaturized components of all kinds, a new type of flat screen—which simply wasn't available in 1949. It is the product of very rapid technological growth, some of which had very little to do with the evolution of industrial design. And there is even a further question, which neither of these very different objects quite resolves. Is a television set something which exists in its own right, as all objects are traditionally supposed to do? Or is it, rather, essentially something which articulates an invisible network of services and is meaningless without them?

The designer of a television set has to think of the market, and this is true of the designers of all consumer products. Industrial design, however lofty its declared intentions, is in industrial societies very much part of the selling process. Car-makers bring in model changes at frequent intervals not because technology forces them, and not from a restless desire to present the public with an improved product (though both of these are subsidiary reasons), but because of the concept of 'planned obsolescence'—change is made for the sake of change, or rather, to persuade the customer to trade in the model he has for a visibly newer one.

Industrial design has also become a major means of building corporate images. At the lowest level, companies have tended to regard the employment of an industrial designer as a kind of status symbol—the way an individual might regard the possession of a thoroughbred horse or a rare breed of dog. It is enough to have a designer on the payroll, without going so far as to use him. But large corporations now use designers in a systematic way, as a means of creating personality not merely for one product, but for a whole range of related products; and not just for products alone but for the organization responsible for making them. The industrial designer sells not merely products but a corporate image. If he is descended from the artisan, the architect and the engineer, he can also trace his ancestry to the itinerant vendor of patent medicines—the moralist is also a huckster.

Ideas of this sort take one a long way beyond the declaration made by a man who was, like Philip Webb, one of the pioneers of the industrial design we know today—W. R. Lethaby. Lethaby wrote, with a touch of condescension, that 'the products of industrial design have their own excellence ... good in secondary order, shapely, smooth, strong, well-fitting, useful; *in fact like the nature of the machine itself.*' The italics are mine. The nature of the machine has changed, and our conception of industrial design has changed with it. The purpose of this book is to draw at least a tentative map of what it has become, while at the same time providing the historical background which will set the present situation in perspective.

PART 1

THE EMERGENCE OF A PROFESSION

Industrial Design in Pre-Industrial Societies

Many of the standard principles of industrial design were known to pre-industrial societies, though not consciously formulated by them. This was true from the very earliest times. If one looks first at the most primitive societies known to us, one sees that their tools are typified both by fitness for use, and by an appreciation of the way a particular problem could be solved in terms of the technology and the materials available. Palaeontologists measure man's evolution partly through the changes to be observed in flint implements discovered on sites all over the world. Probably the earliest of all are those found in the Olduvai Valley Gorge in Tanzania. These (fig. 7) date from some 3 million to 500,000 years ago. Compared to some of the implements which came later, these tools are roughly made, but they nevertheless show a clear understanding of the actual nature of the substance from which they are formed, and of the way in which it can be shaped by flaking. There is a narrow range of types—hand-axes, scrapers and pounders—but each type is already adapted to do a different job. In fact, the whole of industrial design is already there in embryo.

More sophisticated flint tools (fig. 8) show unsurpassed elegance and control of form. But these delicate leaf shapes are not produced simply for the pleasure of it, but because they have proved themselves to be efficient in use. The same basic form is produced in a range of different sizes—the small ones are arrowheads, the larger are used to tip throwing spears, and these weapons are employed according to the kind of game which is being hunted.

Standardization and even a kind of industrial production were understood by the civilizations of the Ancient World, and particularly by the Romans. Much Greek and Roman pottery was made by methods which are recognizably industrial, and which must have involved the intervention of a designer. The Greek potters working in Athens are now chiefly celebrated for the red-figure and black-figure decoration which appears on their best wares, and which now supplies most of the surviving evidence about Greek painting of the Archaic and Classical periods.

Equally interesting, however, in the present context, is the fact that these decorations appeared on a standardized range of shapes, each with its own specialized purpose—the *lekythos* or oil flask, the *kylix* or drinking cup, the *askos* for filling lamps, the *psykter* filled with ice which floated in a larger vessel to cool the liquid. The Romans had a large output of fine red mould-made pottery imitating metalwork. This was widely exported from the centres at which it was made, and of course the fact that the pieces were made in moulds, not thrown, ensured that many items were not merely similar but identical. In fact this whole industry producing Arretine wares (they were named after the town of Arezzo in Tuscany) exists in silent contradiction of romantic ideas about the uniqueness, and therefore the psychological significance, of the pre-industrial hand-made domestic object.

6 (*left*) English folding stool, 15th century. Illustration from *The Industrial Arts* (1876).

7 (*right*) Flint tool from the Olduvai Valley Gorge, Tanzania, dating from 3 million to 500,000 years ago. London, British Museum.

Illustration page 12. Gerrit Rietveld, Zig-Zag chair, 1934.

8 Neolithic stone spearheads, from north-west Australia. London, British Museum.

A more complex example of standardization is Roman weaponry. Rome relied on the power of her armies, and her soldiers were outfitted to a series of standard patterns. The magnificent Praetorian Guard who were the emperor's élite troops did not wear outfits chosen according to their own fancy, but were equipped with identical shields, helmets and swords (fig. 9), produced by slave labour in factories set up for the purpose. Uniformity of weapons and equipment was essential to Roman military tactics, which assumed that a large body of men could be deployed as a single mass.

It is particularly interesting to examine the European Middle Ages for evidence of proto-industrial thinking, since the propagandists for craft as an activity opposed to industry have so consistently discovered their ideal in medieval society. The medieval maker was, on occasion, perfectly capable of the kind of structural logic, economy and ingenuity which we now expect from the best post-Bauhaus furniture designers. One type of medieval folding chair (fig. 6) would look perfectly at home in a showroom displaying modern Danish furniture. But this accidental resemblance is less interesting than some others. In the Middle Ages, as among the Greeks and Romans, there existed a high degree of standardization. Many of the English imperial measures were already fixed at this period—from early in the twelfth century, for example, the English foot was exactly the one now in use, giving three feet to a yard, six to a fathom, and $16\frac{1}{2}$ to a rod, pole or perch. Naturally this affected the shapes and proportions of buildings and the sizes of many standard household articles. The local potteries were highly organized. In England they tended to turn out a standard range of types, the same shape often appearing in a graduated range of sizes (fig. 12). The tile industry was even more highly

9 Roman relief showing soldiers of the Praetorian Guard. Paris, Louvre.

10 Stone mould for casting seals, English, 14th century. London, British Museum

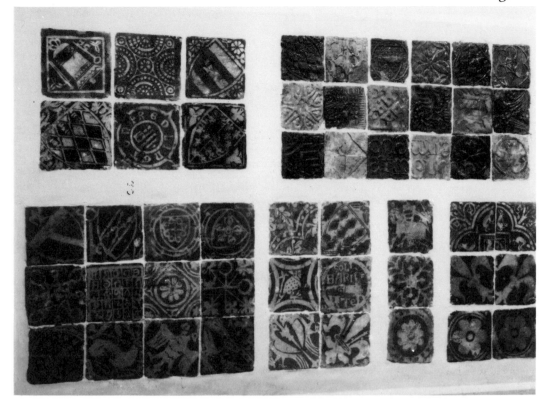

11 Mass-produced tiles, English, 14th century. London, British Museum.

12 Pottery types from Lincoln, 12th-14th century. Illustration from *Medieval Pottery*, Shire Publications (1978).

organized than the potteries which made domestic wares. Though the normal procedure was to build a kiln on site and make tiles there, decorated tiles of better quality were often transported for long distances (fig. 11). In 1396, the tiles needed for Porchester Castle were moved from Billingsgate to the Pool of London, then shipped from the Pool to Porchester itself. Tiles were marked to indicate responsibility for workmanship—at Worcester cathedral all the medieval tiles are marked.

Medieval artisans, like the Roman potters who made Arretine ware, knew the convenience of the casting process when it came to making things in series and at the same time repeating the form exactly, and moulds for making all kinds of objects have survived, among them the mould for making seals which is shown here (fig. 10). In a society which was still partly illiterate seals were of great importance for verifying documents, and it was convenient to have a supply of identical blanks, ready for engraving when either a replacement or a new design was needed. This is in fact a simple example of design logic applied to a particular type of production. Handicraft was applied where it was needed—for the actual engraving—and not for its own sake.

Medieval attitudes towards design were still very much present in the workshops of the seventeenth and eighteenth centuries. It is a temptation, for example, to think of a rushlight holder as a typically 'craft' object, especially when one is confronted with a whole series of slightly differing examples (fig. 14). The temptation is increased because the rushlight itself is no longer part of our own technology, having been replaced by the electric light bulb. However, if one looks at the holder itself without prejudice, one sees immediately that it is in fact very well designed

13 George II red walnut commode/library chest.

14 Wrought iron rush-holders, 18th century. London, Victoria and Albert Museum.

for its purpose, which is to hold a burning rush dipped in tallow in a safe and stable fashion. A modern industrial designer, handed the same problem, would be hard put to find a better basic solution, and rushlight holders in fact vary much less among themselves than the whole vast range of modern table lamps, which are solutions to the problem of holding a bulb so that the light it gives will be effective and fall in the right place.

Eighteenth-century design philosophy, though expressed in different terms to our own, was in many respects very close to that professed today. Designers excelled in devising plain but practical forms, with just enough ornamental detail to prevent dullness. The George II walnut commode illustrated (fig. 13) is basically a plain rectangular box. But its rectangularity is relieved both by the waist moulding under the top drawer and by the bracket feet. These are especially important visually as they link the chest firmly to the ground, without making it seem heavy. Modern designers have been unable to surpass plain Georgian furniture of this type, and indeed many of the new domestic designs we see now contrive to be adaptations of eighteenth-century originals.

This is something which appears clearly if we look, not at furniture, but at Georgian cooking utensils. A set of copper cooking utensils from the kitchen of the first Duke of Wellington at Apsley House is admirably plain and practical in form (fig. 16). Indeed, a number of the same forms can still be found for sale today, though few of us are likely to need a *batterie de cuisine* of some 550 pieces, which is what the Duke considered necessary. It is interesting to compare these saucepans with a series of stainless steel cooking utensils designed by the eminent Scandinavian industrial designer Sigurd Persson in 1978. The shapes are in many ways quite strikingly similar—it is only the

15 Contemporary view of the kitchen, Brighton Pavilion.

16 Part of the Duke of Wellington's *batterie de cuisine*, early 19th century.

17 Stainless steel and aluminium 'Gunda' saucepans and pots, designed by Sigurd Persson, 1978.

18 Chippendale chairs, mid-18th century. Illustration from T. A. Strange, *English Furniture, Woodwork, Decoration . . .* (1890).

actual basic material chosen which is different (fig. 17). A print showing the kitchen of Brighton Pavilion as it was in George IV's day (fig. 15) helps to reinforce the point. The ornate oriental decorations of the public rooms in the pavilion are echoed here by the tin foliage which crowns the cast-iron columns supporting the roof. But the *batterie de cuisine*, though obviously as vast as the one which belonged to the great Duke, is just as plain and practical in form, with skillets and saucepans all of the same shape ranged in a carefully graduated hierarchy of sizes.

Metalwork in precious metals could, by contrast, be extremely ornate, both for reasons of ostentation and to show how much the craftsman-designer appreciated the fine quality of the material he was using. Yet a great deal shows extreme functional simplicity. The first English teapot (fig. 20), which dates to about 1670, is made of silver and looks more like a coffee-pot to twentieth-century eyes. But it shows an admirably direct use of material. A kettle on a stand, of about

21 (*right*) Sheraton chairs, early 19th century. Illustration from Strange, op. cit.

19 English silver tea-kettle on stand, 1710–20.

20 Replica of the first English silver teapot, *c.* 1670.

1710–20, is almost equally plain. It is only in the curving cast feet of the stand that a little Baroque exuberance breaks out (fig. 19).

Eighteenth-century concern with visual style led to the issue of numerous pattern-books for the guidance of furniture-makers and their patrons. Those put out by Thomas Chippendale and George Sheraton are among the most famous. It is only when one examines a range of Chippendale or Sheraton furniture that one begins to realize that the style associated with each

22 (*far left*) Spinning-wheel, mid-18th century. London, Geffrye Museum.

23 (*left*) Mahogany spinning-wheel by John Plants, *c.* 1790. London, Victoria and Albert Museum.

24 (*above*) Spice mill in rosewood, late 18th century. London, Victoria and Albert Museum.

man possesses a strict grammar of its own which is, in turn, based on a certain parsimony of invention (figs. 18 and 21). Sixteen Chippendale chairs are illustrated, for instance, all with different backs. But their legs all follow the same basic form—straight in front, sweeping outward at the rear. In five of these chairs, the stretchers which brace the legs are arranged in an identical way. Three more show only a very slight variation, with the member between the back legs raised a little higher. The legs of the Sheraton chairs also play variations on a very limited number of themes. But it is in the Chippendale chairs in particular that one sees evidence of a firm sense of constructional logic—a firmer logic, indeed, than that shown by some of the founding fathers of modern design, such as Charles Rennie Mackintosh.

It would, however, be idle to pretend that there are no differences between eighteenth-century design attitudes and our own. The eighteenth-century household possessed many fewer machines than a contemporary one, and these machines were often of a type now completely obsolete. Few modern households consider a spinning-wheel to be a necessity. A standard mid-eighteenth-century example (fig. 22) is sturdily constructed of wood, following a design which had evolved over a long period. The turning on the legs and on the spokes of the wheel reveals the maker's love of ornament—something which would be less individually expressed at the present day. A somewhat later spinning-wheel (fig. 23), designed for a more elegant setting, is cleaner in line—but the fact that it is made of fine mahogany, banded with satinwood, reveals its status as a drawing-room ornament. A spice-mill in rosewood, of about the same period (fig. 24), shows the same concern with sturdiness and practicality which one finds in other late Georgian kitchen utensils. Here the precious wood is chosen as much for its extreme hardness as for its looks—the hardness enabling the spices to be crushed without undue wear to the mill itself. The elaborate

25 Silver 'New Universal' microscope, second half of the 18th century. London, Science Museum.

26 Rose engine, German, *c.* 1750. London, Science Museum.

turning which ornaments this handsome piece nevertheless reveals the maker's feeling that both the fine quality of the wood and the preciousness of the spices it was used to crush ought to be celebrated in some way.

Eighteenth-century designers produced a wide range of precision instruments for various purposes, some of which, like the nautical chronometers illustrated in a later chapter (figs. 344, 345), were of considerable complexity. They were sometimes unable to restrain an exuberant feeling for decoration, especially when the instrument in question was produced for an important patron. A German rose-engine of *c.*1750 (fig. 26), belonging to the Science Museum in London, is clearly, like the later of the two spinning-wheels, a drawing-room piece, made for a rich amateur. Even more startling is the ornate microscope in silver produced for King George III (fig. 25). This, with its flower-swagged urns and allegorical figures, is a *tour de force* of the silversmith's art, and in theory of the instrument-maker's also. Yet it is fairly clear that in this case the ornamental features interfere with the efficiency of the instrument. This instrument illustrates the occasional ambiguity of eighteenth-century attitudes towards design, and the relationship of design to function— things which design theorists have since been trying to clear away.

Anyone interested in the pre-history of design must be prepared to look beyond Europe, simply because so many of the leading designers of our own day have drawn inspiration from non-European sources. Islamic art, for example, has been laid under contribution by many leading designers, from the mid-nineteenth century onwards. The powerful forms of some Ottoman metalwork (fig. 27) foreshadow what leading modern designers have tried to achieve, and

27 (*below*) Ottoman silver water-bottle, 16th century, and bowl, 18th century.

28 (*right*) Isnik tile, 16th century.

do it perhaps better than they, because the shapes are less self-conscious. Islamic manipulation of abstract pattern (fig. 28) has been especially influential. The sixteenth-century Isnik tile illustrated treats the problem of repetition without monotony in an especially subtle way. Such tiles were used to cover large areas of wall in mosques and other buildings.

An even more profound contribution to modern design philosophy has been made by the peoples of the Far East. Chinese and especially Japanese tools and implements of all kinds (figs. 30 and 31) seem to have achieved functional perfection through a long period of evolution, without the conscious intervention of a designer. These tools continue to be manufactured in precisely the same form at the present day because nothing better for the intended purpose has been

discovered. Certain of them—the Japanese pull-saw (fig. 31) is a case in point—have become increasingly popular in Europe, as craftsmen discover their superior qualities. It would be hard to discover a better example of Japanese design philosophy than the traditional iron cooking pot (fig. 29) which is the Japanese equivalent of the more familiar Chinese *wok*. As with the *wok*, its rounded bottom enables things to be cooked swiftly, on high heat, with little oil, by the 'stir–fry' process whereby food cut into small pieces of regular size is continually moved around in the hot oil with the help of a pair of cooking chopsticks. Three small legs enable the pot to stand safely and independently on uneven surfaces. The handle can be unhooked and placed in any one of five pairs of different holes in the two lugs on either side— this

29 Traditional Japanese iron cooking pot with wooden lid/stand.

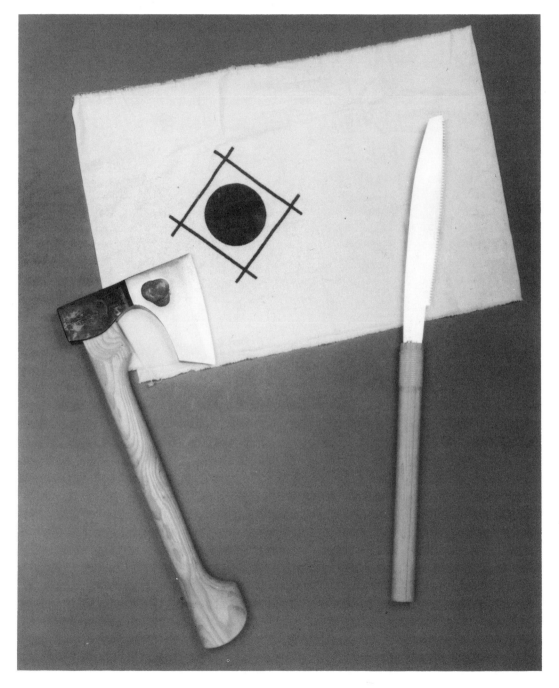

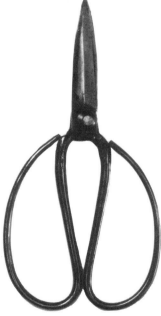

30 Chinese scissors of traditional pattern.

31 (*right*) Traditional Japanese carpenter's axe (left) and pull-saw (right).

32 Engraving of the East India clipper *Bengal*, 1812. London, Science Museum.

33 (*right*) Model of a Viking ship of *c*. 900. London, Science Museum.

34 (*far right*) American Indian birchbark canoe. Dearborn, Michigan, Greenfield Village and the Henry Ford Museum.

means that the vessel can be hung at either a steep or shallow angle if so desired. The wooden cover which fits into the pot can also be used as a stand.

When trying to trace the sources of modern design, one must also be prepared to think in cross-cultural terms, and especially in terms of the kind of problems which recur in many different cultures. Water, demanding and dangerous, as well as being a convenient means of transportation for men and goods, excited the usually anonymous talents of designers in many cultures. Each of the three very different vessels illustrated (figs. 32–34) is a supremely elegant and practical solution to the problem of transport by water. The double-ended birchbark canoe, meant to be paddled, is the simplest. It is very light—the man in it has to move little more than his own weight and anything he may be taking with him. It can easily be lifted out of the water and carried overland until occasion comes to launch it again. It is made of easily available materials. The clinker-built Viking ship is also light in comparison to its size and the number of men it can carry. It can be rowed or sailed, and it can be pulled up, because of its shallow draught, on almost any beach. The Viking raiders and traders who used these ships sometimes sailed immensely long distances in them—as far, on some occasions, as the coast of North America. The design of the longship, however it evolved, was ideally suited to

the demands its users made on it. The East Indiaman *Bengal*, illustrated in a print published in 1812, is an early form of clipper, and represents the penultimate stage in the evolution of the sailing ship as an efficient means of transportation—the final stage being the American tea-clippers (fig. 32) built for the China trade slightly later in the nineteenth century. The aim was to have a fast, weatherly ship, able to transport a reasonable amount of goods reliably over long distances. During the course of centuries of trade with India and the Far East, European shipbuilders tried a number of different forms of ship, among them the huge Portuguese galleons of the late sixteenth and early seventeenth centuries. Often built of teak in Indian shipyards, these immense ships with their high superstructures were too large for the nautical technology of the time, and this unwieldiness, combined with faulty charts and occasional attacks by privateeers and pirates led to frequent losses. The *Bengal* differed in almost every respect, being flush-decked, clean-lined and of moderate size. But she owed something to them, as a sum total of previous maritime experience.

One thing which has been too little noticed, by writers on the history of design, is the way in which, as a conscious design movement began, experience gained in various specialized areas, and especially at sea, began to fertilize the whole design concept.

The Industrial Revolution

Though it is convenient to regard the Industrial Revolution as a unique historical divide, there are in fact several such revolutions in the course of European history. The one to which we accord the dignity of capital letters is only the most recent of the series—or perhaps the penultimate if one takes the effect of the microchip into account. It is important to note that the changes wrought by this eighteenth- and nineteenth-century Industrial Revolution were often even more significant from the sociological than from the purely technological point of view. It gained its main impetus through the textile industry, and its innovations were indeed far-reaching so far as

machines went, but textile design was affected remarkably little. What the new machinery brought with it was a great shift in population from the country to the town, the erection of large new factories (fig. 35), and the accumulation of vast quantities of new wealth, much of it in the hands of people not imbued with the then accepted standards of taste.

It was not the first time that changes in the textile industry had precipitated far-reaching social change. There had already been a crisis in the Middle Ages, set in motion by the gradual emancipation of peasants from the land, the growth of towns, and the labour shortage associated with the Black Death. One response had been a kind of mechanization—machines ranged from things like the ingenious Italian silk-throwing mill shown here (fig. 37) to huge water-powered fulling mills. The overall result was that spinning and weaving and finishing cloth ceased to be something which took place almost entirely within the confines of the family unit, and very largely for the family's own use. During the later Middle Ages cloth was perhaps the major item of European trade, and it was the cloth merchants who pioneered the development of modern capitalism. In medieval England, where wool was the main prop of the nation's economy, the beginnings of modern industrial organization soon started to establish themselves despite the attempts made to check them both by royal authority and by the guilds. One finds not merely the deployment of capital from central sources but division of labour and the bringing together of the workshop under one roof for greater control of the manufacturing process.

The introduction of printing in the fifteenth century marked another significant step towards a fully developed industrialism. Printing put its emphasis squarely on the idea of exact repetition. Once the type was set up the same page could be printed in as many copies as were wanted, merely by putting more paper through the press. All the errors originally made by the compositor would appear in each of these copies if no effort was made to check the result, after which the press could be stopped and the necessary corrections made. Books, once the privilege of the rich, became widely available, and the population was increasingly literate as a result. Availability fuelled demand, and eventually very large printer's shops, like that of the Plantins in Antwerp (fig. 36), were created to supply it. These printer's shops exported their production all over

36 Sixteenth-century printers' workshop. Antwerp, Plantin Museum.

37 Italian silk throwing mill, from a 14th-century manuscript in the Biblioteca Ambrosiana, Milan.

35 (*opposite*) Cotton Mill, New Lanark, 1810–12.

Europe from centres of learning such as Antwerp and Venice. The increasing dissemination of print led to a standardization of available information, and as people became aware of this they also became more systematic about organizing information itself—one of the first prerequisites for real industrial change. Even taste was to some extent standardized by the invention of printing. The birth of the pattern-book, lavishly illustrated with engravings, tended to

deprive the craftsman of a good deal of his creative initiative.

Another industry, too little studied in this context, which early accepted a kind of industrialization was shipbuilding. When one looks at the exuberant decoration our ancestors applied to ships—for example, the rich carvings which adorned the stern of the late seventeenth-century warship *Royal William* (fig. 38)—it is perhaps difficult to accept this view. But if one turns to surviving shipbuilder's plans, the matter appears in a different light. The plan made to guide the builders of the eighteenth-century ship *Atalanta* (fig. 40) is admirably precise, drawn by a man who expected his instructions to be followed to the letter. This precision was necessitated not only by the demands of the sea itself, which tested every design to its limit, but by the large scale of the shipbuilding operations of the period. Contemporary views of the royal naval dockyards at Woolwich (fig. 39) suggest a real parallel to the assembly-line production which was not in fact developed until the early years of our own century.

There is one thing which distinguishes the shipbuilding operations of the period, massive as they were in scale, from what we would now see as true industrialism. This was the continuing dependence on wood as the primary material. Shipbuilding was conservative in this respect, and continued along traditional lines long after the Industrial Revolution had fully established itself elsewhere. One of the key factors in the change was the increasing availability of metal, particularly iron, and its use in bulk. The difference can be observed even in printing, once one of the pioneering spheres but now comparatively conservative. If one compares two presses, one made around 1700 (fig. 41), and the other in 1804 (fig. 42), one notes that they are recognizably the same basic

38 Model of the stern of the 17th-century Warship *Royal William*. Greenwich, National Maritime Museum.

39 Nicolas Pocock, *Royal Dockyards at Woolwich*, 1790 (detail). Greenwich, National Maritime Museum.

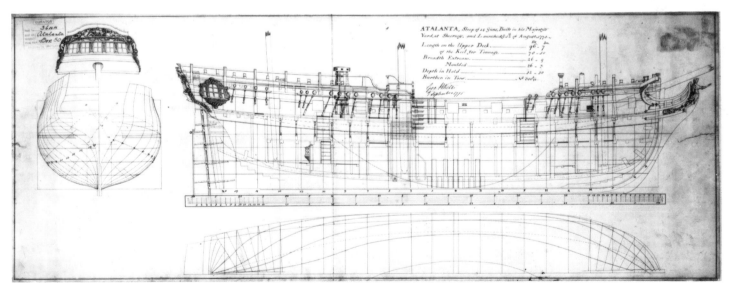

40 Builder's draught of the 14-gun ship *Atalanta*, 1775. Greenwich, National Maritime Museum

machines, despite some refinements added to the later of the two. The biggest difference is to be found in the fact that one is largely made of wood, and the other of metal.

In the mid-eighteenth century the growth of the iron industry transformed parts of the English land-scape. Its altered appearance alarmed contemporaries because they knew that a way of life was being ir-revocably destroyed, but at the same time it stirred their imaginations. Philippe de Loutherbourg's view of *Coalbrookdale by Night* (fig. 43) conveys information about the new iron-foundries through pictorial devices which were already well-established in the seventeenth-century, thanks to painters like Gaspard Dughet and Salvator Rosa. A similar sort of excitement is communicated on a different occasion by the Japanese prints which commemorate Japan's naval successes in the Russo-Japanese war (fig. 44). Here, as

in the painting by Loutherbourg, long-established conventions of representation are reformulated in order to allow the artist to deal with entirely fresh subject-matter—the Japanese printmaker portrays the dreadnoughts of the period using techniques employed earlier to make portraits of the most popular geishas in Edo. These superficially different images have something fundamental in common—they bring us face to face with societies which have been temporarily knocked off balance by the pace of technological change.

The first iron-furnace at Coalbrookdale was set up in 1708. By the third quarter of the century output and quality had both increased to the point where pre-viously undreamed-of things became possible. In 1774 T. F. Pritchard designed and built the first large-scale structure to be completely made of iron—sited in the Coalbrookdale district, it was the rapidly

41 (*far left*) Wooden printing press of *c.* 1700. London, Science Museum.

42 (*left*) Stanhope printing press dated 1804. London, Science Museum.

43 Philippe de Loutherbourg, *Coalbrookdale by Night*, 1801. Oil on canvas, 26¾ × 42 in. (67.9 × 106.7 cm.). London, Science Museum.

44 (*left*) 'Japanese Warship Sighting a Bird'. Print by Terazaki Korgyo from *Pictorial News of the Naval War with Russia*, 1904. 11¼ × 17¼ in. (28.4 × 43.7 cm.).

45 (*above*) The Iron Bridge, Coalbrookdale, 1774.

celebrated bridge at Ironbridge (fig. 45). This structure offers a good example of the way in which the exemplary use of a new material, or of a known material in a totally new context, can alter the whole grammar of design. It was the forerunner of numerous ambitious iron-framed structures built in the succeeding century. A number, such as the great Engine House at Swindon (fig. 46), were built in connection with the new railways. These were themselves examples of the new industrialism, and at the same time a major factor in increasing the pace of its development.

The availability of coal from the Shropshire coalfield made Coalbrookdale attractive to manufacturers of all kinds, among them the makers of ceramics. Pottery was made at Madeley, clay pipes at Brosley, and china at Caughley and Coalport. The original Coalport China Works have recently been restored as part of the Ironbridge Gorge Museum (fig. 47). The scale is surprisingly domestic—more like a farmyard than our current idea of a factory.

Coalport, though sited so advantageously, did not play nearly such a striking role in the history of the Industrial Revolution as another firm of potters—Wedgwood. Josiah Wedgwood (1730–95) founded his new factory in 1769, between Hanley and Newcastle-under-Lyme, at a place he dubbed Etruria. His aim was to convert (and these are the words of his epitaph) 'a rude and inconsiderable manufactory into an elegant and important part of national commerce'. More than any other manufacturer, Wedgwood was connected

with Neoclassicism, the great stylistic change which started to overtake the European decorative arts just after the middle of the eighteenth century. But Neoclassicism involved more than style—it also meant a profound intellectual shift. It went hand in hand with the rising tide of rationalism, with a trust in principles, rules and method. In these circumstances, design could no longer be something haphazard.

Wedgwood quite consciously divided his production into two parts. The most prestigious part consisted of ornamental wares, produced in the bodies he called jasper and basalt. Both were dense, unglazed porcellanous material—basalt was matt black, jasper was made in various solid colours, with a distinctive blue predominating. It was especially well adapted to the imitation of Classical hardstone carvings, since a cameo effect could be obtained by applying moulded ornament in white jasper to a coloured ground. To show the possibilities of his invention Wedgwood produced copies of the Portland Vase, a Roman burial urn of the second century AD made of cameo-cut glass. This was one of the most admired of all known Classical antiquities at that time; and Wedgwood's version, though made of a totally different material, was uncannily faithful to the original. But Wedgwood could never be content with simple copying. He commissioned designs from eminent contemporary artists, chief among them John Flaxman and George Stubbs. These well-known fine artists, without Wedgwood's flair, would probably never have come into contact with the manufacturing process, and they perhaps deserve to be ranked among the first industrial designers.

More significant for the future, however, were Wedgwood's 'useful', as opposed to his 'ornamental' wares. The bulk of these were made in a new kind of fine cream earthenware, subsequently dubbed

46 J. C. Bourne, *The Engine House, Swindon,* 1846. Illustration from J. C. Bourne, *History and Description of the Great Western Railway,* (1846).

47 The restored Coalport China Works, Ironbridge Gorge Museum.

48 (*above*) Transfer-printed Wedgwood Queen's Ware, c. 1770.

49 (*right*) Creamware shapes from Wedgwood's catalogue of 1774.

50 (*above right*) Wedgwood Queen's Ware kitchen ware, c. 1850.

51 (*far right*) Wedgwood teapot designed in 1768 and still in production.

52 (*opposite page*) The first Wedgwood pattern book with border designs for Queen's Ware, 1774–1814. Barlaston, Wedgwood Museum.

Queen's Ware. This was first produced in 1763, several years before Wedgwood moved to his purpose-built establishment at Etruria. Queen's Ware combined quality and cheapness, and the suitability of the body for casting made volume-production possible on a scale previously unknown in the industry. Wedgwood had realized that a new middle-class market existed, unable to afford the fine porcelains produced in various small establishments in England, but anxious for something better than the rough earthenware which seemed the only alternative. Etruria was planned to take full advantage of all the mechanical devices then available, and there was considerable division of labour.

Having invented his new ware, and set up the means of making it in quantity, Wedgwood was not content to rest on his laurels. From 1773 onwards he printed catalogues which were widely distributed, eventually adding French, Dutch and German versions to the English editions. With this aggressive marketing strategy to back it, the reputation of Queen's Ware soon spread all over Europe. Catherine the Great of Russia, always interested in any novelty, ordered a large service for one of her palaces, hand-painted with English views. The majority of Wedgwood's clients were content with less in the way of decoration. Pattern-books were kept, showing standard border patterns which any reasonably skilled artisan could be trained to reproduce (fig. 52). But even this method of decoration was insufficiently rapid. In 1752 the Liverpool firm of Sadler & Green had developed a technique for printing transfers onto pottery, and Wedgwood now hired them to produce patterns to suit his own requirements (fig. 48).

The evolution of Queen's Ware decoration, significant because it became an integral part of the industrial process, is nevertheless not as interesting in the long run as the evolution of the actual shapes used by the factory. Designs from Wedgwood's catalogue of 1774 show traces of mildly Rococo influence, no doubt due to the fact that many items were more or less direct copies from contemporary silver (fig. 49). They may in some cases have been moulded directly from originals in metal. By the early years of the nineteenth century, Queen's Ware forms have for the most part become severely plain. The development was not a consistent progression. In some cases—a familiar teapot is a case in point (fig. 51)—Wedgwood's designers devised completely authoritative shapes at more or less the first attempt. This teapot has been in continuous production since 1768 and is an example of a phenomenon to be discussed later—the way in which a handful of designs leap out of any stylistic category and become timeless.

One great advantage of Queen's Ware was that it could serve many functions. It could appear quite appropriately on an elegant tea-table but it could just as readily be used in the kitchen. It is tempting to associate the magnificently simple kitchenware which the firm of Wedgwood produced in this body in the mid-nineteenth century (fig. 50) with the tradition of stripped-down vernacular design which has been assiduously traced by modern design historians looking for antecedents of twentieth-century functionalism. Yet granted the history of the firm, can anything by Wedgwood be regarded as vernacular in this sense? These nineteenth-century kitchenwares are visibly rooted in the Queen's Ware patterns of some forty to seventy years earlier, and occasionally they seem to reach back very directly to specifically Neoclassical sources of inspiration. In the illustration, the beehive-shaped honey-pot at the lower right is a direct copy of a shape popular with sophisticated late Georgian silversmiths.

The firm of Wedgwood plays an important role in the history of modern design not merely because the first Josiah was a pioneer of rationally organized industrial production, but because he gave rationalism itself a visible, material existence in the shapes his factory produced. Wedgwood provides a living link between the second half of the eighteenth century and our own design revolution.

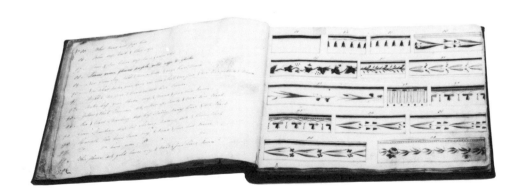

The Suspicion of Industrial Simplicity

The Industrial Revolution triggered off a debate about design which has continued unabated until the present day. It was not long, for example, before intellectuals began to evolve theories of design, and soon thereafter even the politicians started to worry about it. The great Neoclassical architect Sir John Soane was in reaction against the elaborations of late Georgian taste when he declared that 'it is in simplicity that all real decoration is to be found.' His German colleague, Friedrich Weinbrenner, author of a textbook on architecture published in two parts in 1810 and 1819, went even further, actually suggesting that ideas about perfect and beautiful form should be gleaned from the utilitarian objects nearest at hand—things such as drinking-glasses and other domestic utensils. This anticipated by more than 30 years John Ruskin's conclusion that the art of architecture began with the shaping of the cup and the platter. By 1832 the politicians had joined in—it was in that year that Sir Robert Peel publicly blamed falling exports of British goods on incompetent designs.

As early as the 1820s, a great reversal of taste had begun in Britain. People were no longer content with the stark, stripped-down forms of Neoclassicism. They wanted something richer. The reign of George IV was marked by a revival of the eighteenth-century

Rococo style, mingled with lush ornament directly taken from natural forms. A teapot made by a Dublin silversmith in 1828 (fig. 55) is typical of the new style, which was to enjoy a long life despite much public condemnation. It even reappears in the 'naturalistic' porcelains which were being made at Worcester in the 1880s (fig. 54), and can be considered a not-so-remote ancestor of the Art Nouveau of the end of the century.

This style, though it made its appearance at least a decade before Queen Victoria came to the throne, clearly satisfied something deeply rooted in the Victorian psyche. It suggested a kind of succulent abundance which contemporaries found reassuring. For all the abundance of material goods made available by the new machines, the Victorians often felt threatened by the world about them. In England they were without many of the economic safety nets which society now provides—as Dickens's novels demonstrate, it was very easy to tumble from prosperity to ruin.

In the United States, a wholly new society was being built. The American mid-nineteenth-century interior was if possible even more cluttered than its British equivalent. Here the equation was very clear—to

Americans a superfluity of possessions served as a reminder of material success, a reassurance that the days of poverty and struggle were over. 'The things were not status symbols in the modern sense,' writes one American expert on the nineteenth-century decorative arts, 'intended to demonstrate something to neighbours. Although no doubt the adornment of the Victorian house had something—even much—of that purpose, I submit they had more to say to their owners than to neighbours.' The English Victorians, too, had a great deal to survive—not merely the dislocations of the Industrial Revolution itself but the severe post-Napoleonic agricultural depression and the political crisis which had culminated in the Reform Bill of 1832.

These were the subconscious reasons why the Victorians rewarded themselves with material goods. The actual forms taken by those goods were shaped by other factors as well. The eighteenth century had taken great delight in discovering cheap ways of imitating formerly expensive decorative finishes. Sheffield plate imitated the best patterns in silver, scagliola imitated marble, ornaments in patent Coade Stone replaced architectural trimmings carved from

53 (*opposite*) Revolving bookcase. From Tallis's *History and Description of the Crystal Palace*, 1851.

54 (*right*) Worcester cream jug, dated 1881.

55 (*far right*) Silver tea-pot by J. Frey, Dublin, 1828.

the real thing. Thanks to the new machines this tendency started to run riot in the decorative arts. The simplicity of the Neoclassical style had led to a loss of traditional skills in certain trades—woodcarvers practically vanished because Neoclassical designers had preferred smooth surfaces relieved by a little inlay—and this made the short cuts provided by the machine particularly welcome when the fashion for a richer kind of ornamentation was revived. It was also true, though perhaps to a somewhat lesser extent than the propagandists for functional design have imagined, that there was a sort of snobbery involved. The newly prosperous middle class aspired to the luxuries which had once distinguished the aristocratic life-style, and they eagerly purchased objects designed to look as if time and effort had been spent on them—even when this was not in fact the case.

Hatred of machine imitations of handwork was later to become one of the motivating forces of the new design movement. It gave a powerful impetus to the English Arts and Crafts Movement in particular. Yet it was almost as if the Arts and Crafts Movement implanted the love of apparent hand-finish still deeper in the collective middle-class psyche. If one looks for really striking examples of machines imitating hand-finishes, many are belated. The electric kettle designed by Peter Behrens for the German firm of AEG in 1921 (fig. 56) is an elaborate imitation of hand-beaten metal in the Arts and Crafts style of 40 years earlier; and the same style reappears in an English wall-light of the late 1940s (fig. 57), complete with a grease-pan which catches no drips, and smoke-cap hovering over a flame-shaped electric light bulb.

A more immediate and crucial problem in early Victorian design was stylistic eclecticism. No one idiom was dominant. The revived Rococo was challenged by numerous alternatives, all of them derived from the

56 Electric kettle designed for AEG by Peter Behrens, 1910.
57 (*right*) Electric wall-light in current production, 1949.

decorative arts of the past. There was a continuation of the Classical style of the earliest years of the century, and a newer Italianate manner branching away from this. There were also versions of Gothic (which was being taken more and more seriously thanks to the efforts of A. N. W. Pugin, designer of the rich Gothic interiors of the rebuilt Houses of Parliament); plus a kind of Tudor popularized by the novels of Sir Walter Scott and given reality by country-house architects like Anthony Salvin. None of these styles was a middle-class invention—Salvin, for example, was patronized by the grandest aristocratic clients, newly enriched by coal-mines and railways—but they provided the new middle-class clientele with a bewildering variety of role models.

The state of design caused increasing concern to the intellectuals of the day. Some, like Ruskin, denounced the Industrial Revolution and its consequences, and recommended a return to the Middle Ages. Others, like Henry Cole and the group of men who surrounded him, attempted more practical reforms. Cole was a civil servant, the founder in 1849 of a periodical called the *Journal of Design*, edited by his associate Richard Redgrave, a practising designer in his own right, and one of the driving forces behind the Great Exhibition of 1851. Cole's own designs, put out under the trade-name Summerly Art Manufacturers (fig. 58), are certainly not revolutionary. The teapot which is one of the best-known of his creations is clumsy in form, and its ornamentation hovers between naturalistic and Renaissance taste, with an inclination towards the latter — the spout, in particular, seems to derive directly from a sixteenth-century bronze waterspout; while the handle, incorporating a head of Pan, is perhaps taken from a Renaissance lamp.

The Great Exhibition, when it took place, turned

58 Teapot designed for Summerly Art Manufactures by Henry Cole. London, Victoria and Albert Museum.

out to be a curious mixture. The building that housed it — Joseph Paxton's Crystal Palace — was inspired by the greenhouses Paxton had previously built for the Duke of Devonshire at Chatsworth, and derived more remotely from the utilitarian factory buildings of the 1790s, which made extensive use of iron to reduce the risk of fire. Yet it is, as Sir Nikolaus Pevsner points out in his seminal book, *The Sources of Modern Architecture and Design*, 'the mid-nineteenth century touchstone'. Pevsner lists the characteristics which made the Crystal Palace a pointer to the future: 'The Crystal Palace was entirely of iron and glass, it was designed by a non-architect, and it was designed for industrial quantity production of its parts.' He rightly sees it as the direct successor, not only of the structures mentioned above, but of T. F. Pritchard's bridge at Coalbrookdale, which he would in any case prefer to attribute not to Pritchard, who was a minor architect, but to the ironmaster Abraham Darby.

This radical structure aroused less controversy than the exhibits it contained. Some, like an oil-lamp designed by the Frenchman M. Vittoz (fig. 59), took relatively trivial domestic objects and treated them as if they were architectural monuments. The globe of the lamp is borne aloft by a complex structure made of precious metals. It was not a new development to treat things like lamps and clocks as if they were architecture in miniature — it was done very frequently by the creators of the French Empire style of the beginning of the century, men such as Percier and Fontaine. But now the designers seemed to lose all restraint. A ladies' worktable became a coffer in Rococo style, crowned with a group of *putti*, and held up by writhing legs which seemed scarcely able to support its weight (fig. 61). Designers tried to exploit the possibilities offered by all kinds of new materials, the products of Victorian inventiveness and technical ingenuity. Thus a large console-table, the glass surmounting it, and a bracket more or less en suite, were all of them ornamented with gutta-percha (fig. 62). The plasticity of the material was exploited to carry the rococo to even greater heights of extravagance than one finds in the worktable already cited. A monumental cabinet, somewhat loosely based on a French late Renaissance original (fig. 63), was designed to display the merits of a new and equally novel ornamental process. The wood was not carved, but the rich decoration was burnt into it, so as to imitate carving.

Other items of furniture showed a search for novelty of form as well as of ornamentation. One such (fig. 53) was a drum-shaped bookcase revolving on a horizontal shaft, so that a series of pivoted shelves successively presented themselves at the right height to the user. Cylindrical bookcases revolving on a vertical shaft had been fashionable during the Regency, and this was a variation of the same principle, with many vestigial details to link it to the late Regency style, for example the paterae and other classical details on the side-members and the lion's-paw feet. But here again one is conscious that an apparently trivial item is aspiring to inappropriately monumental form.

More genuinely radical in its conception was the metal-framed, sprung, revolving chair exhibited by the American Chair Company in New York (fig. 64)—one of several chairs with frames wholly or largely of cast-iron, steel, or some combination of the two. Metal furniture was not wholly unknown previously—the Spanish had made metal stools in the seventeenth century, and metal had also been used for the campaign furniture popular during the Napoleonic Wars—but one does seem to see here an effort to rethink the basics of furniture construction. However,

59 Oil-lamp in gold and silver manufactured by M. Vittoz and shown at the Great Exhibition of 1851, from the *History and Description of the Crystal Palace and the Exhibition of the World's Industry*, 1851.

60 Moulded glass imitating cut glass. 1950s?

61 Worktable, from the Great Exhibition catalogue.

62 Gutta-percha console table and frame, from the Great Exhibition catalogue.

63 Renaissance-style cabinet, the wood burnt to imitate carving, from the Great Exhibition catalogue.

64 Revolving chair, by the American Chair Company, from the Great Exhibition catalogue.

the American designer did not have quite the courage of his own functional convictions, since the metal legs which support the stem which in turn supports the spring are elaborately scrolled.

Though official catalogues praised both the ingenuity and richness of the exhibits, and though the general public which flocked to the exhibition was impressed by them as evidence of material progress, they triggered a sharply negative reaction among the inner group who had promoted the whole thing in the first place. Richard Redgrave, editor of the *Journal of Design*, which Cole had founded, wrote a *Supplementary Report on Design in the Great Exhibition*, which restated the functionalist doctrines that had already been propounded by one or two Neoclassical theorists. Writing of china and glass, Redgrave said, 'the purest forms should be sought, allied to the greatest convenience and capaciousness; and the requisite means of lifting, holding, supporting, of filling, emptying and cleansing, should engage the attention of the designer, before the subject of their ornamentation is at all entered on.' This post-Great Exhibition outcry was to have a long echo—perhaps

because the would-be reformers signally failed to secure what they had in mind. As late as 1937 we find Pevsner writing, in his despairing *Enquiry into Industrial Art in Britain*: 'A pressed-glass bowl trying to look like crystal, a machine-made coal-scuttle trying to look hand-beaten, machine-made mouldings on furniture, a tricky device to make an electric fire look like a flickering coke fire—a metal bed masquerading as wood—all that is immoral. So are sham materials and sham technique' (fig. 60). It is interesting to note that the young Pevsner inherited not only the same cause, but an equally nineteenth-century terminology. It was John Ruskin above all who turned bad design, and particularly the imitation of one material by another which was either cheaper, easier to work, or perhaps both, into a sin rather than a simple error or a lapse from good sense and decorum.

It may seem a little late to begin to defend at least some aspects of Victorian design. Before I even begin to do so it is necessary to say something about the nature of the Great Exhibition itself. Because it was a unique event, and hugely publicized as such, manufacturers were tempted to show, not articles of everyday production, but exhibition pieces specially prepared for the occasion. These were intended to summarize the prevailing technology, to bring new techniques to the attention of the public, and generally to publicize all kinds of possibilities and availabilities. This is the case with nearly all the exhibits selected for comment here. The manufacturers' laudable intentions, translated into domestic terms—that is, applied to objects which were supposedly of everyday use—did produce results which were ridiculously overblown, but the everyday objects in use in Victorian households were not like these, though not all of them were successful designs

by any means. What we also tend to forget is the durable quality of the designs which the Victorians made for public, rather than private, use. It is only recently, for instance, that the quality of a great deal of Victorian street furniture has come to be appreciated at its true worth. Even now, that appreciation remains specialized, and has not yet been fed into general histories of design. In designs for public situations, the Victorian liking for ornament was combined with a sense of dignity and fitness.

Before Victoria's reign urban street furniture was in general very simple, like the handsome cast-iron bollard decorated with the monogram of her prede-

65 Cast-iron bollard with monogram of William IV, 1836–7. London, Trafalgar Square.

66 Victorian pillar box designed for Birmingham.

67 Pillar box of 1860, Lichfield Street, Wolverhampton.

68 Pillar box of *c.* 1866–76.

69 Modern pillar box.

70 Lamp-post on Skeldergate
Bridge, York, 1878–81.

71 Late Victorian Lamp-post
with griffin supports.

72 Cast-iron Hindu well at Stoke Row, Oxon., designed by A. E.
Reade in 1863.

73 Victorian cast-iron and wood public bench, London.

cessor William IV which stands at the base of the statue of King Charles I in Trafalgar Square (fig. 65). Victorian innovations, chief among them gas lighting and the penny post, brought with them need for a much wider range. It is interesting, for example, to follow the evolution of the familiar pillar-box for posting letters. An early example, designed for Birmingham *c*. 1856 (fig. 66), is clearly derived from the bollards which had preceded it. The fluted base explains its name, and it is topped by a dome on which there sits a cushion with a crown resting on it, to symbolize the fact that the post was an official service.

About four years later (fig. 67) we meet a new design—more capacious, and in proportion more like the pillar-boxes which have become familiar. The crown is replaced by a simple royal monogram, V. R. The box, as if to compensate for this loss, is lavishly ornamented with baroque garlands and ribbons in low relief. Still later—around 1866—came a much plainer hexagonal type, where the ornamentation is largely confined to the top, which has the kind of finial you might find on top of a teapot (fig. 68). In 1876, the cylindrical pillar-box became standard, and is only now in turn being replaced by something which quite recognizably belongs to our own century (fig. 69). Victorian pillar-boxes, even of the earliest designs, are nevertheless likely to be around for a long time to come. They are physically extremely durable, and they have achieved a total acceptance which makes them functional even today.

Victorian street lighting has had to be adapted to modern requirements, through conversion from gas to electricity, and much has fallen victim to modern planners. Examples which survive run the gamut of possible styles—the handsome gothic triple standards of 1878–81 on Skeldergate Bridge in York (fig. 70), the baroque standards in the grander parts of Central

London, and a number of more eccentric patterns; one, with winged griffins (fig. 71), is yet another version of Renaissance forms, of the kind apparently preferred by Henry Cole. These lamps are extremely successful in a symbolic sense, whatever their practical drawbacks. They put across the notion that the urban environment is something to respect and be proud of, and do it with tremendous panache. The same is true of numerous other Victorian additions to the urban scene—some mildly eccentric, like the Hindu well erected in Stoke Row, Oxon., at the expense of an Indian maharaja (fig. 72), some perhaps slightly monstrous, such as the cast-iron camel which supports a bench on the Victoria Embankment in London (fig. 74), and some by now merely commonplace, like the much more ordinary type of Victorian public bench (fig. 73). All of these are undeniably in the same idiom (or collection of idioms) as the designs shown at the Great Exhibition, but here the contemporary design vocabulary is used successfully, with a keen awareness of what ornament could contribute to the urban landscape.

74 Cast-iron camel supporting a bench on the Victoria Embankment, London, 1870.

75 (*opposite page*) The evolution of the Coca-Cola bottle, 1894–1957.

An Industrial Vernacular

As design history has developed as a separate subject of study, design historians have had a tendency to look for an independent tradition developing naturally within industry, in response to the needs of the mass market which the machine had created. These historians see the new industrial vernacular as something which is made up of various elements. There is, for example, a kind of *ad hoc* practicality—designs which combine pre-existing elements so as to produce something original and new. An example, though it was not something subsequently manufactured in quantity, is the first traffic signal, which was installed in Bridge Street, Westminster, in 1868 (fig. 78). It consisted of a cast-iron lamp-post, adorned with predominantly gothic ornament, combined with a railway signal. The mechanism was quite simple—it was operated by a policeman using a lever—and the whole thing, so a House of Commons select committee was told, was the brain-child of one Mr Knight, superintendent of the South Eastern Railway.

The industrial vernacular first expressed itself in recognizable form in the buildings put up to house the machines which were turning out an ever-increasing flow of products. Some (fig. 77) were original chiefly for negative qualities—their utter plainness and lack of adornment, their acceptance of a purely utilitarian role. Others, like the striking factory buildings at Noisiel in France, erected circa 1874 (fig. 76), showed an unprejudiced attitude towards materials. The buildings are iron-framed, an extension of the principles pioneered by Paxton.

Some items escaped adornment because they were not considered visually prominent or important enough to merit it. The sturdy mid-Victorian brass coat-hooks which still survive in considerable quantity (fig. 79) have a simplicity of form which indicates quite accurately that they were below the notice of either snobs on the one hand or intellectuals with theories about design on the other. The sturdy glassware used in Munich wine restaurants in the mid-nineteenth century (fig. 80) also excites admiration today because of its unadorned practicality. A number of the standard items used in these establishments, particularly the stemmed wine-glasses with their conical bowls, are adaptations of traditional designs which had become popular a century previously. Munich tavern glass owed its simplicity to a number of different factors. First, it had to stand up to hard use—the restaurants were crowded and quite often rowdy. Secondly, the actual nature of these establishments—unpretentious and popular—made simplicity socially appropriate. Thirdly, they were made for an essentially conservative clientele—both the restaurant owners and their customers wanted shapes which seemed familiar.

There was also something more to it—the German-speaking world, and more particularly Austria and Southern Germany, had an independent design tradition which often expressed itself in a preference for very simple forms. In furniture, the Biedermeier style took over some ideas from the Napoleonic period in France. The massively simple architectural shapes were domesticated and made simpler still by the absence of trim in gilt metal. The same sensibility expressed itself in ordinary domestic objects and utensils—for instance, in a stoneware coffee-pot of c. 1820 which is the German answer to Wedgwood's Queen's Ware (fig. 81). The regional tradition persisted into the mid-century. Even factories which concentrated on the richly decorated glass which had now become popular, for instance, would nevertheless produce conspicuously plain wares for those who thought differently. An example is the so-called Musselinglass, designed by Ludwig Lobmeyr of Vienna in 1856. These glasses (fig. 82) are not designed for tavern use—in fact the bowls of the glasses are conspicuously thin, which is the reason for the name, and the stems, where present, have a slenderness to match. The design found immediate acceptance with the public, and is still being produced at the present day. Its attraction lies not only in the simplicity of form, but in the way in which the container shows off the liquid it contains—these are glasses for the connoisseur of fine wines, and therefore appeal to a conservative clientele as much as they do to an avant-garde one.

The culmination of the Biedermeier tradition of furniture-making, and at the same time a conspicuous departure from it if looked at from a technical point of view, was the bentwood furniture produced by the Viennese firm of Thonet. Michael Thonet began by experimenting with veneers, which were laminated together and then bent to shape (fig. 83). He thus anticipated experiments made in our own century by Marcel Breuer, Alvar Aalto and others. Thonet's first chairs were comparatively laborious and expensive to make, and it was the much simpler bentwood furniture which made his fortune. Thonet bentwood varied from the drastically simple (fig. 84), to larger pieces whose exuberant curlicues expressed a

76 Cast-iron factory building at Noisiel, France.

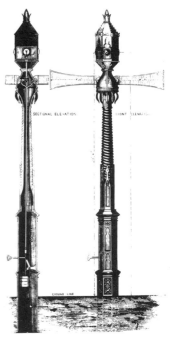

78 Traffic signals, Bridge Street, Westminster, 1868. Illustration from *The Engineer*.

77 (*left*) New Ing Mills, 1863.

79 (*above*) English mid-19th-century cast-brass hooks. London, Victoria and Albert Museum.

80 Glassware in use in 19th-century Munich wine restaurants.

81 German stoneware coffee-pot made at Zell-am-Hammersbach, *c*. 1820. Karlsruhe, Badisches Landesmuseum.

82 Musselinglass, designed by Ludwig Lobmeyr, 1856.

decorative energy from which not even he was immune (fig. 85). The secret of Thonet furniture lay not only in the originality of the way in which the necessary shapes were created, but in the logic with which the whole process of manufacture was organized. The various parts were screwed together rather than glued (which meant that the furniture could be shipped flat). Seats, leg-braces and other parts were interchangeable and were used in designs which were apparently very different from one another.

The Biedermeier tradition was important historically. It created the particular cast of mind which made it possible for the Bauhaus to happen, and Bauhaus designs, as they recede from us in time, begin to look more and more like Biedermeier brought up to date, just as the architecture of its last director, Mies van der Rohe, more and more obviously reveals his debt to the great German Neoclassicist Schinkel. But it was a development which remained isolated within Central Europe.

If one wants to make out the case for the development of a true industrial vernacular, one has to turn one's eyes towards Britain and the United States. There were certain areas of design where the nineteenth century never completely lost its taste for simplicity. Horse-drawn vehicles, for example, had an aesthetic of their own, which existed in striking contradiction to the aesthetic of the contemporary domestic interior. (One reason for the contrast may have been that there was a distinction between masculine and feminine taste. Carriages were the province of the former, and drawing-room furniture of the latter.) The gentleman's travelling carriage of

83 (*far left*) The first bent veneer chair made by Michael Thonet, 1836–40.

84 (*left*) Thonet 'Vienna' bentwood side chair with plywood seat, 1876.

85 (*above*) Thonet bentwood chaise-longue, late 19th century.

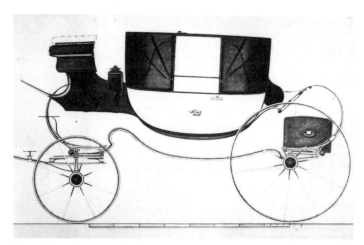

86   Early 19th-century light coach.

87   American 19th-century racing sulky.

the early years of the century (fig. 86) was a design which had been refined by successive generations of coachbuilders over the best part of a century. Within the terms set—a horse-drawn vehicle to carry passengers in comfort over long distances and still fairly rough roads, and at reasonably high speed—it was scarcely possible to find a better solution. Where the designers of the later years of the century did provide some new answers was in devising horse-drawn vehicles which were solutions to specific problems. An obvious example was the racing sulky (fig. 87), a skeleton frame with a seat, supported on a single pair of large wheels. The racing sulky was a development of the light trotting sulky (fig. 88), which seated two people side by side and which was popular in America for its minimal elegance. The hansom cab (fig. 89) was an English invention—a light vehicle plying for hire in the streets of London. It superseded a much clumsier four-wheeled type known as 'growlers' from the notorious surliness of

the coachmen who drove them. The hansom, with its two wheels, was an ingenious solution to the problem of carrying either one or two passengers in comfort and privacy. The driver was perched high up at the back, which kept him out of the passengers' way and at the same time gave him maximum visibility.

These horse-drawn vehicles, like the graceful clipper-ship which was more or less contemporary with them (fig. 90), represent the end of something, not the beginning of something. The automobiles and steamships which succeeded them in consequence of the inexorable series of technological developments triggered off by the Industrial Revolution were not at first nearly so logically or economically designed for their particular roles. Because they did not lead forward directly, it is hard to see sulkies and clipper-ships as the indubitable ancestors of contemporary industrial design, though efforts have been frequently made to prove that this is what they are. Perhaps the strongest justification for such a theory is to be

88 American 19th-century trotting sulky by C. W. Watson.

89 London Hansom cab, *c.* 1900.

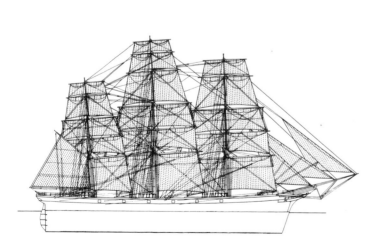

90 Drawing of a clipper ship, *c.* 1865. London, Science Museum.

found in the writings of the American sculptor Horatio Greenough, whose essay *Form and Function*, written in 1851, was one of the first attempts to put forward a coherent plea for functionalism. It was Greenough who wrote: 'The men who have reduced locomotion to its simplest elements, in the trotting wagon and the yacht *America*, are nearer to Athens at this moment than they who would bend the Greek temple to every use.' But Greenough's writings caused little stir at the time when they were first published.

Those who have investigated the idea that a kind of pragmatic functionalism spontaneously manifested itself in the Victorian epoch have also paid a good deal of attention to so-called 'patent' furniture. This was not in fact something completely new. As early as the late sixteenth century, furniture designers had started to devise items to meet special needs, and especially the needs of invalids. A drawing survives depicting a chair made for Philip II of Spain, who like many of his contemporaries was a sufferer from gout. It can be seen that the angle of both the back and the footrest were adjustable by means of metal ratchets (fig. 91), and it also looks as if the arms were hinged to drop outwards. Probably, like much Spanish furniture of the period, it was designed to fold up altogether for travelling. In any case, it is undoubtedly the direct ancestor of the Victorian adjustable invalid chair which is also illustrated (fig. 92), though here the framework is made entirely of metal. The fact that metal is used

throughout is not so much a measure of technological progress in this instance, as of social psychology—the chair had its own specific purpose, and was not required to fit in with decorative norms. Iron bedsteads (preferred as more hygienic and unlikely to shelter bugs) benefited from the same tacit exception to normally observed rules.

The Victorians, with their admiration for mechanical ingenuity of all kinds, were fascinated with the idea of furniture which could be transformed in

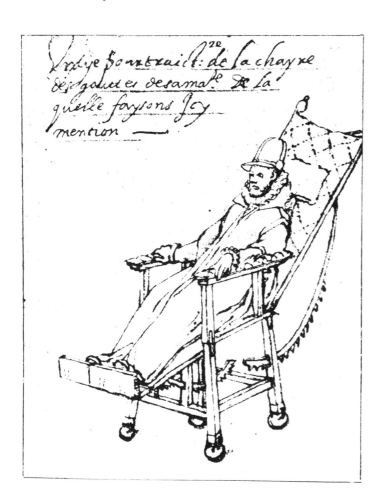

91 (*left*) Invalid Chair made for Philip II of Spain, 16th century. Illustration from *Dictionnaire d'Ameublement*, vol. III.

92 (*above*) Victorian adjustable invalid chair. High Wycombe, Parker-Knoll Collection.

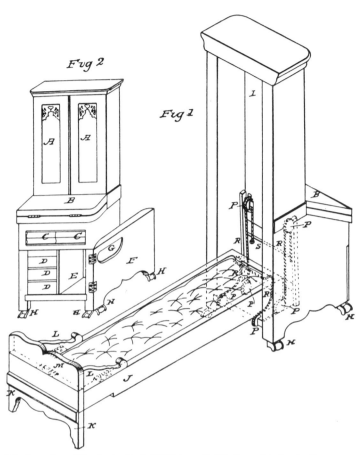

93 American Wardrobe Bed, 1859 From S. Giedion, *Mechanisation Takes Command*, 1948.

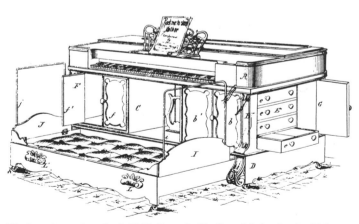

94 American piano bed, 1866. From S. Giedion, *Mechanisation Takes Command*, 1948.

surprising ways. There were beds which dropped out of wardrobes (with a chest of drawers clapped on to the back, fig. 93), and there was also a piano which turned into a bed, and which also included space for bedclothes and a washbasin, plus a built-in bureau (fig. 94). It is significant that such pieces are known to us now largely through diagrams and patent specifications—it is unlikely that they were manufactured in any quantity. In fact, it is considerably easier to find actual specimens of mechanical furniture made before the Victorian epoch.

In the eighteenth century, furniture designers were torn between two conflicting impulses. There was a multiplication of types—for example little tables for every imaginable specialist purpose: work-tables, tables incorporating a firescreen (so a lady could sit close to the fire to write a letter without burning her face), tables for a gentleman's dressing-room where he could empty his pockets before retiring for the night. There was also an interest in what was called 'combination furniture'—pieces which had dual or triple functions. English designers, such as Sheraton and Hepplewhite, included designs for such pieces in their pattern-books, concentrating on dressing-tables and writing-tables (fig. 95). The interest in dual-purpose furniture was increased at this time by the housing shortage brought about by the Napoleonic Wars. Despite the fact that they seemed to espouse diametrically opposite views of the way in which furniture should function, the specialist items and the dual-purpose ones did have something very much in common. Both were apt to incorporate ingenious mechanisms, and these became the more ingenious the more luxurious the piece.

The Victorians, inevitably, continued this tradition of mechanical furniture, and there has been a tendency amongst historians of industrial design to claim these items for their own subject. In fact this claim is more than doubtful.

A more valid claim for a place in the pantheon of early vernacular design can be made for other specialist furnishings—those created especially for travellers. Ship's fittings were designed to make the maximum use of the space available, and voyaging Victorians, like their eighteenth-century predecessors, often took specially made furniture with them. They used it during the voyage; then, if they were emigrating to a new country, it provided a basis for furnishing their new homes. By its very nature furniture of this type did mark the beginning of a distinctively functionalist tradition.

Britain's colonies also made their own distinctive contribution. In India, the custom of going on safari led to the production of furniture specially adapted for the needs of the traveller, and much of it was produced by Indian craftsmen. An example is an ingenious collapsible chair made of wood, canvas and leather straps. Disassembled, this can be carried in a simple canvas bag (fig. 98). It was such a successful design that it began to be produced in Britain as well, from 1907 onwards, and was still on the English market in the early 1950s.

One of the most ingenious of these designs for travellers is not generally thought of as being 'furniture' at all—I mean the massive wardrobe trunks favoured by prosperous voyagers of a rather later epoch. The Harrods catalogue of 1914 illustrates a number of splendid specimens, and Harrods continued to offer these trunks long after the First World War—almost identical models are to be found in their catalogue of 1928–9. Such contraptions, with their ingenious fittings (fig. 97), took the place of the ship's furniture used earlier. Though the exterior finish is different, they are in precisely the same tradition of design. They obviously inspired the wardrobe designed, in the late 20s, by a much respected avant-garde figure, Eileen Gray (fig. 96).

95 Marquetry coiffeuse made by Abraham Roentgen for Friedrich Augustus III, Elector of Saxony. Dated 1769.

96 (*below*) Prototype wardrobe designed by Eileen Gray and Jean Badovici, 1926–9. Paris, Musée des Arts Decoratifs.

97 (*above*) Secretaire trunk designed by Louis Vuitton for the conductor Leopold Stokowski, 1936.

98 (*left*) Hardwood and canvas collapsible chair of Indian design, first produced in Britain in 1907, and still in production in 1956.

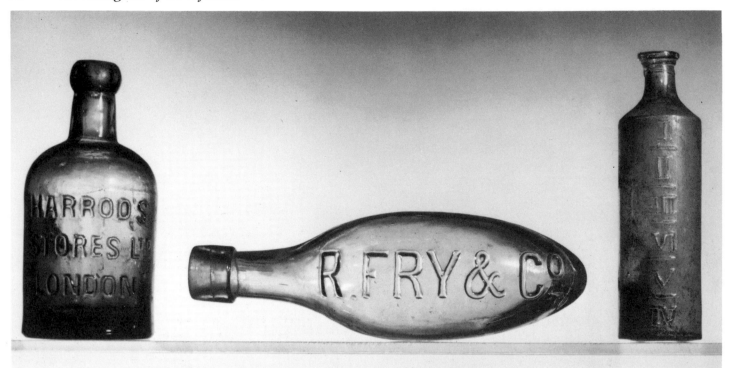

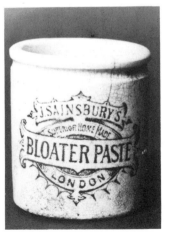

99 (*top*) Nineteenth-century soft drink bottles.

100 (*above*) Doulton storage vessels and containers, redesigned in 1948.

101 (*above right*) Early 19th-century snuff container in lead and earthenware.

102 (*right*) Sainsbury bloater paste jar, 1920s.

103 (*opposite*) Stages in the evolution of the modern jam-jar.

If there is an important and direct connection between the vernacular design of the Victorian epoch and the more self-conscious and carefully calculated industrial design of the present, it is to be found, not in furniture, but in the glass and pottery containers made for commercially produced food and drink. It was in fact the increasingly professional marketing of familiar products which inevitably brought changes in the way they were presented. The Fribourg & Treyer earthenware snuff-jar of George IV's reign illustrated here (fig. 101) is an example of commercial packaging at an early stage of its development. Jars of this form, or something like it, were basically derived from the apothecary jars of the fifteenth century, and they were used as containers for materials of all kinds (fig. 102). They have recently been revived, as part of the contemporary cult of Victorian packaging—a

theme I will return to later—and are, even more interestingly, the ancestors of the modern glass jam-jar (fig. 103), surely an ultimate example of industrially inspired directness and lack of frills. Soft drinks also inspired a wide range of designs for bottles—in the mid and late nineteenth century these were made of thick glass with raised lettering (fig. 99). Coca-Cola began its triumphant commercial career in just this kind of bottle, at the very end of the century, and the bottles it is sold in now are still visibly linked to nineteenth-century prototypes (fig. 75). Nineteenth-century storage vessels and containers have stood the test of time so well that designers seem to be able to do very little with the basic forms. For example, when Doulton redesigned their range in 1948, they still came up with shapes which would have seemed at home in any nineteenth-century kitchen (fig. 100).

An Age of Invention

The nineteenth century added new types of material to those which were known already, and it also added wholly new types of object. Celluloid, the first plastic, was invented in 1865. Like other new materials, it caused the Victorian designer small difficulty. In his mind these materials fell into an already established category of imitations and substitutions. A celluloid hairpin-box, dated 1888, and therefore one of the earliest datable celluloid objects (fig. 105), is an imitation of the same kind of article made of pressed horn, a material quite familiar at the period. It was only gradually that designers began to discover that the new materials possessed positive qualities of their own.

In many cases, the Victorians, when confronted with some new invention, tried to find the means of bringing it into line with what they knew already. This was especially true if it had to be fitted into the domestic environment. A gas fitting of 1859, intended for use in dining-rooms (fig. 106), derives its basic shape from the lanterns containing oil-lamps which had been in use 50 years earlier under the Regency, and from candle-lanterns of an earlier period still. It was only raw laboratory experiments, like Alexander Graham Bell's first telephone (fig. 107) of 1876, which were left crudely unadorned, and this crudity was soon modified as soon as there was any question of their straying into the house.

Victorian inventions can be categorized in different ways. There were those, for example, which seemed to take a decisive step forward, achieving their definitive form comparatively easily. Examples which come to mind are the safety bicycle and the fountain pen. There is a world of difference, for example, between one of the first fountain pens (fig. 110) and the quill pens (fig. 108) or even those with steel nibs which had preceded them. But there is not very much practical difference between this early fountain pen and the sleek Parker 51 introduced in 1939 (fig. 109), though the Parker 51 has been redesigned to fit in with the then-dominant 'streamline style' which affected many other domestic objects during the 1930s, as well as the trains, ships, automobiles and buses to which it was more logically applied.

The safety bicycle had a slightly more complex evolution. It was preceded by the hobby-horse (fig. 104), which the rider kicked along with his feet, and by the penny-farthing (fig. 111), with its large

105 Celluloid hairpin box, made in Antwerp, 1888.

104 (*left*) A 'Dandy-horse' bicycle.

106 (*right*) Ornamental gas fitting, originally published in *The Universal Decorator*, 1859.

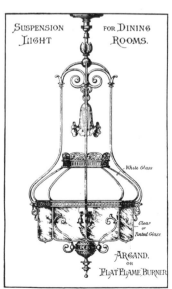

SUSPENSION LIGHT   FOR DINING ROOMS.

White Glass

Clear or Tinted Glass

ARGAND. OR FLAT FLAME BURNER

Type of Ornamental Gas Fitting by Sugg in use about 1875.

107 Alexander Graham Bell's first telephone, 1876.

wheel driven directly by the pedals. The penny-farthing was too precarious a mount, and required too much agility and athleticism, to make bicycling available to everybody. This was remedied by the introduction of the chain-driven safety bicycle in the 1870s (fig. 112). Essentially the bicycle then took on the form which it has retained ever since. A Singer Safety Bicycle of 1890 (fig. 113) looks very like the models we are familiar with today, and a Raleigh bicycle of 1905 (fig. 114) is even more convincingly modern. The bicycle seemed to accept certain modifications fairly easily, while rejecting others which were on the face of it less radical. Thus, it

adapted itself quite readily to the advent of the internal combustion engine (fig. 115), but failed to absorb an innovation proposed in the early 1920s—the chainless bicycle (fig. 116). The strength of the design was its structural logic and lack of complication, and it is these qualities which have ensured its durability.

Another innovation which at first sight falls into the same category is the electric light bulb. There is little apparent difference between Edison's experimental carbon-filament lamp of 1879 (fig. 117) and commercially produced Ediswan lamps of *c*. 1890 (fig. 118). Both are the direct and close ancestors of the light bulbs we use now, though the contours of the modern product have been smoothed by mass-production, and sockets have been modified for greater convenience and security. What the light bulb brought with it, however, was a whole new range of problems connected with the way it was to be shaded and supported. Hanging lamps and table-lamps alike became the subject of intensive design investigation—an investigation which continues unabated. Unlike the single unitary solution to a problem of locomotion

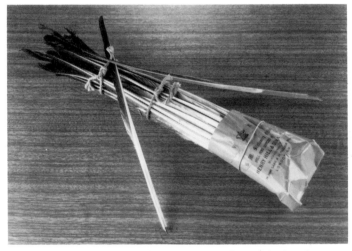

108 Bundle of quill pens, as they were originally supplied to the British civil service. London, Science Museum.

109 Parker 51 fountain pen, designed 1939.

110 The Onoto self-filling ebonite fountain pen, 1890s.

111 Penny-farthing cycle, 1879.

112 The first chain-driven safety cycle, 1873–4. London, Science Museum.

113 Singer safety bicycle, 1890. London, Science Museum.

114 Raleigh bicycle, 1905.

115 Holden motorcycle, 1897. London, Science Museum.

116 F. N. chainless bicycle, *c.* 1920. London, Science Museum.

induced to reproduce herself directly, without man having to labour to trace her lineaments, appealed deeply to the Victorian scientific spirit. In fact, in a curious way it seemed to link Victorian science to Victorian religion. God recorded his creation directly, through the action of light on a chemically sensitized surface. The English photographic inventor Fox Talbot, who was responsible for introducing the calotype (the first negative to positive photographic process), described how his use of the camera obscura as an aid to making drawings led directly to his successful photographic experiments: 'the idea occurred to me—how charming it would be if it were possible to cause these natural images to imprint themselves durably and remain fixed upon the paper!' The title he chose for the book in which he gave an account of his invention was *The Pencil of Nature* (1844).

117 Edison's first carbon filament lamp, 1879. London, Science Museum.

118 Commercial Ediswan lamp, *c.* 1890. London, Science Museum.

which was offered by the safety bicycle, here there were a multitude of possible answers, and lamp-design became a complex dialogue between functional efficiency on the one hand, and aesthetic and cultural considerations on the other.

People did at least have an idea already present in their minds about the way in which they expected a table-lamp to look—and this was the designer's opportunity as well as a limitation imposed from outside. The same could not be said about the camera. Photography was one of the most significant of Victorian inventions. When the first really practicable process was announced in 1839, the public had long been eager for it. The idea that nature could be

119 The Kodak roll-film box-camera, 1888. Harrow, Kodak Museum.

120 Wet-plate camera on stand, *c.* 1865. Harrow, Kodak Museum.

121 Daguerrotype camera, 1839. Harrow, Kodak Museum.

Designing a camera, which was essentially a light-tight box equipped with a lens, was something which fell well within the capacities of nineteenth-century instrument-makers. It was not the camera itself which made early photography complex, but the instability of photographic chemicals and the need to use them freshly mixed. A wet-plate camera of the 1860s (fig. 120), designed to record the desired image on a glass negative from which an unlimited number of positives could be made, shows little advance on the early daguerrotype camera (fig. 121), which used to make unique positives on specially prepared metal plates. The wet-plate process was the one used by the majority of the great Victorian photographers—it superseded Fox Talbot's method, which produced paper negatives. Camera design began to change only when the gelatine process, which took over from wet-plate photography, was married to celluloid—this made possible the introduction of commercially prepared roll-film, and turned photography into an art anyone could master. The Kodak box camera of 1886 (fig. 119) marked the beginning of a new generation of camera design.

But the situation was not as simple as this brief account might suggest. Beautiful craftsman-made cameras of wood, metal and optical glass continued to be constructed for special purposes, and indeed are still obtainable new today. The camera, once the basic principle had established itself, spanned any number of variations, but it was not until very recently that this principle—that of light coming through a lens and falling on a chemically sensitized surface—started to be challenged by new introductions like the Sony Mavica (fig. 357), which records the image on a video cassette so that it can be replayed through a television set.

Also typical of Victorian inventiveness were a spate of new office machines. As business increased its tempo, new devices were brought in to lighten the load. A Burroughs adding machine of 1898 (fig. 122) represents the business technology of the last decade of the century. It is not powered by electricity, and of course owes nothing to the microchip, yet its design from a purely mechanical point of view is most elegantly simple and functional. The manufacturer shows his pride in the solutions he has found by providing glass panels through which the various moving parts can be seen. The machine is light in weight and unornamented apart from the trademark on the front panel. Closer examination shows how carefully designed it is from the user's point of view. The keyboard is inclined, to put all the keys at the same distance from the eye; the keys themselves are removable for repair, and the whole mechanism is easily accessible to a mechanic should repairs be needed. The most significant thing about this adding machine, however, is the way it created a format which remained valid until the coming of the miniaturized electronic calculator. Later versions have a similar arrangement

of keys, with the low-value ones to the right of the keyboard, in close proximity to the operating lever.

An even more significant innovation was the typewriter. A Sholes & Glidden typewriter by Remington & Sons of 1874 (fig. 123) has a mechanism which is in some respects differently arranged from that of a modern mechanical typewriter, but the basic keyboard is the same, with a few minor differences— C and X are transposed, and M has risen from the bottom line to the middle one. The floral decoration stencilled onto the metal is significant. It resembles the decoration on early sewing-machines and it tells us that typewriters were, from the first, intended for the use of women. The increasing importance of female labour in offices during the late Victorian period was almost as powerful an agent for social change as the mobility conferred by new means of transportation—the railway for long distances, and the bicycle for shorter ones.

Unlike the camera, the typewriter continued to be subject to a good deal of mechanical experimentation, and various solutions were found to the problems it posed. The Barlock No. 7 typewriter of 1889 (fig. 124) has a double row of keys. The maker still thought it obligatory to ornament his design—the

122 Burroughs sterling adding machine, 1898.

123 Sholes & Glidden typewriter by Remington & Sons, 1874. London, Science Museum.

124 Barlock No. 7 typewriter, 1889. London, Science Museum.

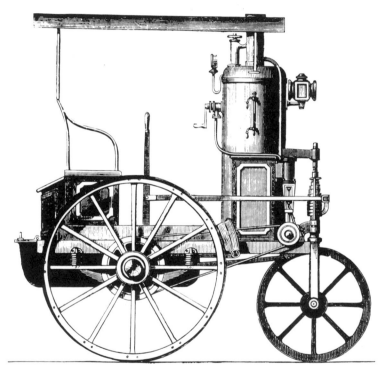

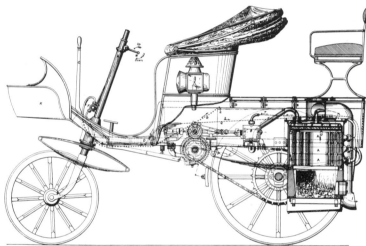

125 (*top*) Traction Engine harnessed to an omnibus, 1871. Illustration from *The Engineer*.

126 (*left*) One-horse road steamer to carry two persons, 1869. Illustration from *The Engineer*.

127 (*above*) Serpollet Steam Phaethon, 1891. Illustration from *The Engineer*.

shield protecting the keys has rococo ornament in low relief—and the lettering is in a style which reflects the influence of the then-prevalent cult of things Japanese stimulated by the Aesthetic Movement.

Of all Victorian inventions, the one which was slowest to settle down into a now-recognizable form was the automobile. One reason for this was the fact that men's ideas about transportation by road were firmly set in a mould provided by horse-drawn vehicles. The other was the competition between the internal-combustion engine and the steam power

familiar since the beginning of the century. Steam was tried on the roads, as well as on the tracks the railroad builders provided. A traction engine hitched to an omnibus (fig. 125) of 1871 is a crude attempt to transfer the railway train directly to the road. A 'One Horse Steamer to Carry Two Persons' of the same epoch (fig. 126) shows a clear derivation from both the railway engine and the traction engine. Two decades later designers were trying a different tack. The Serpollet Steam Phaethon of 1891 is a horse-drawn phaethon with no horses to pull it—and with a steam-boiler tucked under the dickey-seat traditionally provided for the groom (fig. 127). Almost any traditionally trained coachbuilder could have constructed the chassis. The same applies to the very similar vehicles which were produced at the same period, and powered not by steam but by gasoline (fig. 200). It was only gradually that the internal-combustion engine triumphed, and equally gradually that independently powered vehicles took on forms which derived logically from the mechanical necessities on the one hand, and consumer demand for increased comfort and convenience on the other. The change may have been slow not merely because people were so used to horse-drawn conveyances that they found it hard to imagine new and radically different forms, but also because horse-drawn vehicles were themselves such beautiful and economical solutions to certain problems that builders were reluctant to abandon them. We can see the same reluctance at work during the transition from sail to steam in ships.

The railways altered the landscape, and the very act of constructing them also consolidated, if it did not wholly create, a new kind of industrial organization—one on a fully nationwide basis. Labour was deployed on a massive scale (fig. 130), as it had been previously in order to build the canal network which preceded the railway as a means of bulk transportation. What the automobile did was something different. Once the problem of finding a viable form had been worked out, the new vehicle was ready to be democratized and put in the hands of large numbers of people. This event took place in America before it happened in Europe, and it was closely connected not only with the evolution of the automobile itself but with the construction of a system of roads upon which it could travel. The vehicles could be made in sufficient quantity, and then sold at a price which brought them within reach of a broad mass of consumers only by means of a revolution within the factory—a fully developed assembly-line system was introduced (fig. 128). The rows of identical Model-T Fords lined up at the Ford factory in 1914 were eloquent of a major

128 Model-T Fords being assembled, 1914.

129 Pottery ready to go into the kiln at the Copeland factory.

130 J. C. Bourne, *Building the Stationary Engine House, Camden Town*, 1839. Illustration from J. C. Bourne, *History and Description of the Great Western Railway* (1846).

change in design thinking. Industry had long ago discovered how to turn out simple objects which were identical to one another—as I have already pointed out, the secret was known to the ancient Roman potters who made Arretine wares. The stacks of plates and bowls ready to go into the kiln at the Copeland factory (fig. 129) make the point graphically, simply because the picture shows something which was being done in the same way long before the camera was available to record the process.

But an automobile was not something simple and unitary. To design the necessary parts, and then to put together those parts in precisely the same way on each occasion, called for a degree of organization which had never been needed, or even thought of,

previously, not even in the big shipyards, where many fewer units were made. At first the engineers were content to control the whole thing themselves— Henry Ford in his epoch-making early years wanted no truck with stylists, and indeed was content to make occasional changes to the Model-T in reluctant response to technological change, without direct reference to the market, which seemed content to accept what he chose to give it. Yet eventually the industrial designer came to be seen as an essential part of the whole process of manufacture, and also as someone quite different from the engineer, with whom his relations remained ambiguous. Some of the reasons for this ambiguity will emerge in succeeding chapters.

The First Industrial Designer

Christopher Dresser, who was born in Glasgow in 1834, and who died in 1904, is the man who has the best claim to be described as the first industrial designer, or at least the first to be conscious of his role. Significantly, however, Dresser is associated solely with domestic items, not with the products of heavy industry. Whereas the designers who had preceded him fell into three categories—they were architects, amateurs who made their designs *ad hoc*, or artisans and engineers turned designers as a result of practical experience in the workshop—Dresser received a much more academic training, of a kind then just becoming available. He studied at the Government

School of Design at Somerset House, London, from 1847 to 1854. Through his training he came into contact with the circle of mid-Victorian design-reformers headed by Henry Cole and Richard Redgrave. His training spanned both the Great Exhibition of 1851, and the epoch of heart-searching which followed.

There were other significant aspects of Dresser's education. He had a strongly scientific bent, and studied as a botanist, writing books and papers on this subject. His accumulated scientific publications won him the degree of Doctor of Philosophy, awarded *in absentia* by Jena University in 1860. In the same year he was made Professor of Botany at the Ladies' School and the London Hospital Medical School, Professor of Botany applied to the Fine Arts in the Department of Science and Art, South Kensington, and Professor of Scientific Botany at the Royal Polytechnic Institution. He was elected a Fellow of the Edinburgh Botanical Society and was an unsuccessful candidate for the Chair of Botany at University College, London. His scientific studies led to an interest in the relationship between natural forms and ornament—this was the subject of his first important series of articles, published in the *Art Journal* of 1857. In a more general sense, they clearly pointed him towards a rational and logical approach to practical problems of design.

Where ornament was concerned, Dresser opposed the then-flourishing 'naturalistic' school. For him, plant forms had to be conventionalized in order to be useful to the designer: 'Conventionalized plants will be found to be nature delineated in her purest form, hence they are not imitations but are the embodiments in form of the mental idea of the perfect plant' (see fig. 133). But botany, where Dresser was concerned, was more than simply a source of shapes and patterns. In his own phrase, plants demonstrated 'fitness for purpose', or 'adaptation'. He was thus linked, from an intellectual point of view, with early nineteenth-century utilitarianism. Darwin was Dresser's contemporary, and announced his theory of natural selection in 1859, when Dresser was beginning his career. Though the latter apparently stopped short of embracing Darwin's ideas when they were first announced, they certainly influenced him in the long run.

131 Christopher Dresser, electroplated sugar bowl, c. 1886.

From 1862 onwards Dresser's practice as a freelance designer started to blossom. It was in this year that he published his first book on design, *The Art of Decorative Design*. By 1871 he was able to boast, to a critic of his paper, 'Ornamentation Considered as High Art', which he had read to the Royal Society of Arts: 'As an architect I have as much work as many of my fellows, as an ornamentalist, I have much the largest practice in the United Kingdom—there is not a branch of manufacture that I do not regularly design patterns for, and I hold regular appointments as "art adviser" and "chief designer" to several of our largest art-manufacturing firms.' His business interests eventually expanded beyond this. In 1876 and 1877 he paid an extensive vist to Japan, and made a large collection of Japanese objects, some of which were later sold through the firm of Tiffany in New York. In 1879 he entered into partnership with Charles Holmes of Bradford, later the founder of the *Studio* magazine. They had a wholesale warehouse which imported oriental goods. When this partnership came to an end, Dresser was already involved in a new venture—the Art Furnishers' Alliance, founded in 1880 'for the purpose of supplying all kinds of artistic house-furnishing material, including furniture, carpets, wall-decorations, hangings, pottery, table-glass, silversmiths' wares, hardware and whatever is necessary to our household requirements'. The venture was not a financial success, but it was recognized at the time as something pioneering because it tried to reach a popular audience in a way which had not been attempted before. The one self-imposed restriction, and this was a significant one, was that implied by the repeated use of words such as 'artistic' and 'art-manufacture'. The cultivated middle class was attempting to find a practical way of instructing those less fortunate than itself, but still with a determination not to modify its own standards. In late Victorian England there had grown up a distinction in people's minds between furniture and decorative art which responded to these standards, and the rest of the huge output of manufactured goods. Art-manufacture consciously addressed itself to a minority, even though the goods which went under this label were made by precisely the same methods as others which were considered to be beyond the pale. This points to a weakness which continued to dog the profession of industrial design long after Dresser's day.

Dresser's own surviving designs cover a wide range of materials, styles and techniques. He worked, for instance, for the Coalbrookdale Company, making designs for domestic items in cast iron. These date from the 1870s (figs. 136 and 137), and are in a stylized gothic manner which is extremely evocative of the period. Dresser also made designs for glass, and a large number for ceramics. He worked briefly for Wedgwood, and did a much larger series of designs for Minton—the connection lasted from about 1867 until the 1880s. A big collection of his watercolour designs can be found in the Minton archives (fig. 132), and a number of Minton pieces decorated with these survive. They illustrate one of the perils which still lies in wait for the freelance designer—some of the designs are used in the way Dresser clearly intended, as his drawings show, but others are applied to totally unsuitable shapes.

He had better luck with the Linthorpe Art Pottery, founded in 1879 chiefly as a vehicle for Dresser's ideas. At Linthorpe, factory production methods were used—the pottery was inexpensive, and was manu-factured on a large scale. The emphasis was on original shapes, rather than elaborate surface decoration. Dresser turned for inspiration to all kinds of historical

132 Christopher Dresser, design for a vase. Minton archives.
133 (*right*) Christopher Dresser, design for a frieze, 1874–6.

134 (*above*) Christopher Dresser, claret jug and glasses, in glass mounted in silver, designed for Hukin & Heath, Birmingham, 1882. London, Victoria and Albert Museum.

135 (*right*) Christopher Dresser, electroplated teapot, *c.* 1881.

136 (*far right*) Christopher Dresser, cast-iron hat and umbrella stand, *c.* 1875.

137 (*opposite*) Christopher Dresser, cast-iron umbrella stand, *c.*1875.

sources—Pre-Columbian pottery, as well as Chinese and Japanese ceramics. Some pieces even look as if they were inspired by the Minoan civilization which was then still undiscovered, and may indeed be based on Helladic and Mycenaean wares.

Dresser's most original work was in metal, and was produced for various leading firms of Birmingham silversmiths, prominent among them J. W. Hukin and J. T. Heath, and Messrs Elkington & Co. These designs are notable for their simplicity and their direct use of materials (fig. 134). In addition, they often show great originality of form, with strong emphasis on a kind of stripped-down geometric purity (fig. 132). Dresser was one of the first to analyse the relationships between form and function in a rational way. In his *Principles of Decorative Design* (1873) he provided diagrams demonstrating the laws which governed the efficient functioning of handles and spouts on jugs and other vessels, such as teapots. His own teapots (fig. 135) are often extremely distinctive in shape, with emphatic slanted handles. The ergonomic and the metaphorical aspects are skilfully combined. 'I have availed myself of those forms', Dresser wrote, 'to be

seen in certain bones of birds which are associated with the origins of flight, and which give us an impression of great strength, as well as those observable in the propelling fins of certain species of fish.'

Dresser's metalwork also shows his concern with economical use of materials. A plain oval sugar bowl has its edges rolled inward to strengthen the metal at the rim, so that a thinner gauge can be used. Very often, and indeed almost invariably in larger pieces such as soup tureens, Dresser used electroplate rather than silver. This was not a reluctant compromise, as it became with other designers, but a deliberate choice, meant to put his wares within the financial reach of as many customers as possible. His liking for economy expressed itself visually in a famous toast-rack in which the slices of toast are held in place by simple uprights which pass through a metal plate to serve as legs. In these designs Dresser seems to anticipate the Bauhaus. He anticipates it, but he is not a direct ancestor. It is Dresser's surprising success in building relationships with industry as it then existed which seems in some ways to isolate him from the mainstream of orthodox design history.

Design Idealized

There is a paradox in the fact that it has become customary to trace the beginnings of industrial design as we now know it to the English Arts and Crafts Movement. That movement sprang, after all, from a profound revulsion against industry and the society as well as the environment which industry was creating. The precursor of the Arts and Crafts aesthetic was A. N. W. Pugin, who was born in 1812, and who died comparatively young in 1852, only a year after the watershed of the Great Exhibition. Pugin was the great early nineteenth-century partisan of the gothic. Horace Walpole, the promoter of the first Gothic Revival, saw the style as a diversion, and talked of his sham-gothic castle at Strawberry Hill as a 'plaything house'. Pugin, a Catholic convert, treated the gothic style with a deep moral seriousness not untypical of the generation he belonged to, whatever its religious beliefs. He saw it as the characteristic expression of Christian as opposed to pagan culture. Pugin hated the substitution of one material for another, something as characteristic of the late eighteenth century as it was of the early part of the nineteenth; for him things must be made of what they seemed to be made of. He also detested furniture and bibelots which were architecture in little: 'staircase turrets for inkstands, monumental crosses for light-shades, gable-ends hung on handles for door-porters, and four doorways and a cluster of pillars to support a French lamp'. He was an active furniture designer, and the furniture he produced towards the end of his life stressed simplicity and sturdiness in an almost exaggerated way.

Pugin preached, and also taught through his own example as a designer—Ruskin merely preached. But his was an eloquent and highly influential voice. His most effective manifesto was a chapter entitled 'On the Nature of Gothic' which appears in the second book of *The Stones of Venice*, published in 1853. The chapter was reprinted as a separate pamphlet in 1854, and reprinted frequently again thereafter. Ruskin was against the smooth perfection of finish which Neoclassicism aspired to, and which (it seemed) industry was increasingly equipped to provide: 'Men were not intended to work with the accuracy of tools, to be precise and perfect in all their actions. If you would have that precision out of them, and make their fingers measure degrees like cogwheels, and their arms strike curves like compasses, you must un-humanise them.' He stood sufficiently far from factory operations not to be aware of the fact that the designer's task was frequently to find some camouflage for the imperfections inseparable from mass-production.

Ruskin laid down certain rules for craft which were hugely influential. These are the three that probably had most effect:

1. Never encourage the manufacture of any article not absolutely necessary, in the production of which *Invention* has no share.

2. Never demand an exact finish for its own sake, but only for some practical or noble end.

3. Never encourage imitation or copying of any kind, except for the sake of preserving records of great works.

Ruskin's most immediate impact was on a small group of Oxford undergraduates, among them Edward Burne-Jones, the Pre-Raphaelite painter-to-be, and his friend William Morris. Morris, too, had aspirations to be a fine artist, but these were never really fulfilled. His career was made as a writer and publicist, and as the founder of Morris & Co., otherwise called the Firm by those connected with it.

In 1862 a second great International Exhibition was held at South Kensington, and it was here that Morris & Co. made its official début, showing stained glass, embroideries, furniture, table-glass and candlesticks, none of them truly 'industrial' in nature. But Morris did not fear and detest the machine as Ruskin did. Indeed, as he grew older he increasingly felt that it could have virtues of its own if only it were to be properly used. He was at all times obsessed with the idea that the designer must have a thorough understanding of both processes and materials, preferably from his own direct experience, and should strive to make every object the expression of these. Morris found a reservoir of techniques and also of forms in the surviving country workshops, and many of his firm's best-sellers were adaptations of simple, traditional country designs (fig. 141). In any case, this fitted in with his own penchant for simplicity. At the end of his life Morris told Edward Carpenter that he would himself prefer to live with 'the simplest white-washed walls and wooden chairs and tables', rather than with the elaborate wallpapers and rich woven and printed stuffs that were by this time indissolubly associated with his name.

138 Josef Hoffmann, adjustable chair in beech, metal and plywood, *c.* 1905.

139 C. R. Mackintosh, the
Conference Room at the
Glasgow School of Art, 1907–9.

140 C. R. Mackintosh, design
for a dining-room in the house
of an art-lover. Illustration
from the *Zeitschrift für
Innendekoration*, 1901.

Morris set two opposing impulses in motion. One, more immediately influential, was the desire felt by many educated men of his day to reform their own lives, and at the same time transform the state of the decorative arts by finding an alternative to industry; the other and more practical was the feeling that the revived crafts could provide industry with new and better models.

141 The 'Sussex' chair, designed by Dante Gabriel Rossetti and manufactured by William Morris & Co., 1866.

The move to set the crafts up in opposition to industry failed, as it inevitably had to. Charles Robert Ashbee's Guild of Handicraft, founded in the late 1880s, when the Arts and Crafts Movement was at its height, supplies a case in point. It began at Essex House, a Georgian mansion stranded in east London, at Mile End, with subsidiary retail premises in fashionable Brook Street, and for a while it was much in vogue. In 1902 Ashbee transferred its operations and the whole workforce to Chipping Campden in Gloucestershire, with the aim of founding an ideal community. But this cut the Guild off from its market, which remained an urban and luxury one. Ashbee found his organization was unable to compete, not so much with industry as with do-it-yourself—the amateur craftsman or craftswoman who preferred to make craft rather than to buy craft. Ashbee sarcastically and collectively dubbed these amateurs 'dear Emily'. Another, and even more formidable, competitor was the antiques trade, which supplied the same buyers with the mellow products of the past. By 1908 the Guild was in such low water that it had to be dissolved and reconstructed on a far more modest scale. Ashbee himself began to see that there were after all virtues in industry; he now condemned what he called the 'intellectual Ludditism' of Ruskin and Morris.

Though he was himself a talented and original silversmith (fig. 142), Ashbee had little personal influence on the progress of design as a whole. Such influence as he did have came not from the activities of the Guild which he founded, but from his promotion of the work of Frank Lloyd Wright in Europe. In the British Isles, during the late nineteenth century, the one truly original force was the Scottish architect Charles Rennie Mackintosh, now generally recognized as the only major British exponent of Art Nouveau.

Mackintosh and his associates—his wife Margaret Macdonald, her sister Frances and Frances's husband Herbert McNair—made their public début in a completely Arts and Crafts context, at an exhibition organized by the Arts and Crafts Exhibiting Society in London in 1896. This first appearance was not particularly well received. However, Mackintosh did receive a series of commissions in Glasgow—one, an important building which is still intact, was for the Glasgow College of Art. Completely in the tradition of the great architects of the eighteenth century, but in a very different and highly individual style, Mackin-

tosh fitted and furnished various important rooms in this to the last detail (fig. 139). More important from the point of view of design history, however, are the series of interiors he devised for his most important Glasgow patron, Miss Cranston, who ran a series of tea-rooms. The way in which Mackintosh fitted these up, and the striking furniture he devised gave Miss Cranston's tea-rooms a commercial 'livery' which a modern industrial designer would recognize as being completely akin to the kind of thing he is now called upon to produce himself.

Apart from this, Mackintosh's designs were not

142 C. R. Ashbee, glass and silver claret jug (1897) and bowl (1899).

143 (*opposite*) C. R. Mackintosh, chair, 1902–3, designed for Hill House, Helensburgh.

always well conceived. His famous chairs (fig. 143) are usually uncomfortable to sit in, and often show weaknesses of actual construction. There is no technical originality in the way they are made. Yet it is important that Mackintosh made a considerable impact abroad. One major success was the second prize he took in a competition organized by Alexander Koch, an art-promoter in Darmstadt, for a design for a house for an imaginary art patron. This design was published in the *Zeitschrift für Innendekoration* in 1901 (fig. 140), and brought Mackintosh's work to the attention of the avant-garde in Germany and Austria. He was already known to the artists, designers and architects of the Vienna Secession, since just previous to this burst of publicity in Germany he had been invited to exhibit at one of their shows. He and his associates also showed in Venice, Turin, Budapest and Dresden, and on each occasion roused a great deal of interest. England, as opposed to Scotland, was the one place where they failed to make a real success. His designs were important not for their connection with industry, but for their emphatic break with nineteenth-century historicism.

The Vienna Secession, a group of forward-looking artists and architects led by Josef Hoffmann, Koloman Moser and Joseph Olbrich, had been founded in 1897, and was one of a number of such 'dissident' art movements stirring in Europe at this time. Deeply imbued with the ruling Symbolist aesthetic, it was also affected by the British Arts and Crafts Movement— they knew Ashbee's work as well as that of Mackintosh. In 1903 Hoffmann and Moser set up a craft workshop of their own, the Wiener Werkstätte. They were inspired by Ashbee's Guild of Handicraft, and also took advice from Mackintosh. Much of the furniture made in the Werkstätte is spectacularly luxurious—an example is the smoker's cabinet

144 (*right*) School of Hoffmann, side chair in bentwood and plywood, before 1910.

145 (*below*) Josef Hoffmann, basket in metal, painted white, 1905.

146 (*far right*) Sir Ambrose Heal, wardrobe in unpolished chestnut, c. 1905.

147 (*below right*) From a catalogue of Heal & Son, 1898.

NOTICE.

REDUCTION OF PRICES.

In consequence of the immense popularity which the innovation of their New Hygienic Wood Bedsteads has acquired, HEAL AND SON have obtained increased facilities for producing these Bedsteads, not only in much larger quantities, but at a very low cost, and have therefore been able to reduce the prices to their customers from 12 per cent. to 20 per cent. less than their former Price Lists, which are cancelled by this one.

ILLUSTRATED
PRICE LIST

OF HEAL AND SON'S
NEW HYGIENIC WOOD
BEDSTEADS FITTED
WITH IRON LATH OR
STEEL SPRING BOTTOMS

DECORATIVE
HYGIENIC
INEXPENSIVE

entirely faced with mother-of-pearl which Hoffman designed *c.*1910. This has massively simple forms which seem to anticipate the post-war Art Deco of Parisian furniture-makers. Some products of this group of Viennese designers were genuinely radical. This is true of Hoffmann's metalwork (fig. 145), which seems to take Dresser's way of thinking a stage further. It is also true of his adjustable chair (fig. 138) in bentwood and plywood, designed *c.*1905, and of a bentwood and plywood single chair of *c.* 1910 which is perhaps not his but which is closely connected with his school (fig. 144). It is this single chair in particular which gives us a clue to the special nature of Secession design—the imported Arts and Crafts tradition has here started to intermarry with the genuinely local one founded by Thonet over half a century previously.

There was another way in which the Arts and Crafts Movement affected domestic design in general, and this was through straightforward commercialization. Though, as Ashbee discovered to his cost, the economics of true craftwork on the pattern recommended by Ruskin were so unfavourable that any organization founded on Ruskin's principles had a desperate struggle to survive, the *look* of craft—as opposed to its reality—did make an appeal to an educated segment of the middle class. Some consequences of this were grotesque and long-lasting—they are summed up in the candle-lamp (fig. 57) illustrated in a previous chapter, and in a number of other objects shown in this book. But not all the results of the new fashion were so disastrous. A number of commercial furniture-makers in both Britain and America tried to satisfy the demand for what was really a new kind of 'art-furniture'. In America the fashion for the Arts and Crafts look was commercially exploited by Gustav Stickley; in England there were Liberty & Co. and Ambrose Heal. Heal's designs of the

148 Brooklyn Bridge, 1869–83.

149 Adler and Sullivan, Rothschild Store, Chicago, 1890s.

1890s and early 1900s have been described in a recent and authoritative book on twentieth-century furniture, as 'eminently practical' but also 'depressingly spartan'. In fact, the best of his work has a real claim to be considered the real origin of much of the functional furniture which has been issued since (fig. 146). Heal's work was completely straightforward, technically sound and direct, and available to the public at a price almost anyone could afford. The publicity issued by the firm put great stress on practicality linked to economy (fig. 147). It was no accident that these designs were the direct ancestors of the ranges of Utility furniture (fig. 199) manufactured in Britain in the years of stringency during and just after the Second World War.

In a very real sense, the whole Arts and Crafts Movement was a manifestation of a growing consci-

ousness, among educated people, of design and its possibilities. Beginning in Britain, in the aftermath of the Great Exhibition of 1851, this consciousness spread across the Channel to continental Europe, and over the Atlantic to the United States. But it was a consciousness which set its own very strict limits. It was based on a reaction against industry, not on a determination to use it, and this is why a figure like Christopher Dresser is almost unique. Even Dresser seems to have seen his role as being to tame industry and palliate its consequences.

One way of measuring the division between the educated design thinking of the period and its physical context is to look at the schizophrenia which overtook late nineteenth- and early twentieth-century architecture—something different from the stylistic confusion which had existed earlier. Certain structures—the great bridges for example (fig. 148)—remained firmly within the territory of the engineer. Give or take a little superficial ornament here and there, the design was organic, and grew directly from the problem to be solved. Other constructions, such as the skyscrapers and other tall urban buildings going up in America, owed much to the engineer but often borrowed unsuitable traditional forms in which to clothe themselves. They were not, strictly speaking, purely industrial buildings, and contemporaries did not see them as such, but in many technical respects they were unlike any buildings which had been put up before. Architects had to find an idiom to suit them, and their success was intermittent. Some of the most convincing answers to the questions posed by these buildings were discovered in the new cities of the American Middle West, especially Chicago, by men like Louis Sullivan and his peers (fig. 149).

At almost the same time, architects in the Old World were having to find ways to accommodate changes not so much in building technology as in society. The English Garden City movement, for instance, is about the life-style of a new kind of English middle class. In London's Hampstead Garden Suburb the greatest British architect of the period apart from Mackintosh, Edwin Lutyens, supplied some attractive and convincing solutions to the immediate problem. But Lutyens used his dazzling skills to disguise innovation, not to stress it. He adroitly camouflaged new social patterns in what were at first sight reassuringly familiar forms (fig. 150). It was houses like those he designed for the Garden Suburb that Ambrose Heal's bedroom suites would have been used to furnish. Like Heal, and in even more developed form, Lutyens had a gift for seizing the essence of tradition and presenting it stripped of anything extraneous. But the need to find a new industrial idiom was discreetly sidestepped. In the last resort industry remained the enemy of good taste, something not to be used but to be outmanoeuvred. This was something which was soon to change, and very drastically. A rebellion was on its way against all previously accepted standards, and it was to manifest itself first in the fine arts.

150 Sir Edwin Lutyens, houses on the west side of Erskine Hill, Hampstead Garden Suburb, under construction *c.* 1908.

The Revolution in the Fine Arts

The late nineteenth-century decorative arts, more particularly those which addressed themselves to an educated and cultivated public, existed within the broad context of the Symbolist movement. Symbolism had its roots in literature, but came to affect all forms of artistic expression. General currency was first given to the term by the minor French poet Jean Moréas, in a manifesto published in the French newspaper *Le Figaro* in 1886. Symbolism in its first phase involved a dandified revolt against materialism, a retreat into the ivory tower. Industry, therefore, stood for everything which the Symbolists most detested. And this was one of the things which helped to build a bridge between Symbolism and the Arts and Crafts Movement, which was originally quite separate from it. But there was something else as well. For the committed Symbolist, everything had the power to convey hermetic meanings. J. K. Huysmans's novel *A Rebours (Against Nature)*, published in 1889 and one of the key texts of the new movement, is almost as much a manual of interior decoration as it is a work of fiction. There was a reaction, at least for a while, from the division, firmly established since the Renaissance, between the fine artist on the one hand, and the craftsman or decorative artist on the other. The typical Symbolist interior was very much all of a piece, with paintings, sculptures and domestic artifacts making similar and interchangeable statements. The one condition was that nothing must look mass-produced (even if it was). That would have fatally disturbed the necessary feeling of mystery and of special meanings known only to a select few.

But Symbolism, like all major cultural movements, had an inexorable dynamism of its own. Artists and craftsmen who pursued ever more esoteric and refined effects and sensations eventually reached the point where both they and their audience began to feel permanently jaded. The first stage of the reaction is contained within the general current of Symbolism itself, and is summed up in the bold neo-primitivism of Gauguin. But the search for the barbarous soon proved to be as disillusioning as all the other quests the Symbolists had pursued, and eventually a new generation began to feel that there was something even more fascinatingly brutal in the heart of their own society—the machine.

The first group actually to proclaim this view were the Italian Futurists, and it was they who established mechanical objects and the products of industry as key subjects in modern art. Their determination to do so is roundly stated in the famous sentence from the First Futurist Manifesto, published in *Le Figaro* in 1909 : 'A racing automobile is more beautiful than the Victory of Samothrace.' This was an unqualified assertion that the machine could be judged on the same footing as all the other products of man's creativity.

In their paintings the Futurists wanted to render the dynamism of contemporary life—the movements of crowds in cities, and the rapid motion of an automobile or a train (fig. 151). The conventions they devised to render mechanical movements owed much to Victorian experiments with simultaneous photography, especially those made by the Frenchman E. J. Marey. It was a clear example of art being directly influenced by the new technology. The Futurists, however, were not always quite as modern as they made themselves out to be. Essentially their reactions to industry were as superficially romantic as those of de Loutherbourg, when he painted the furnaces of

151 (*opposite*) Gino Severini, *The Train in the City*, 1914. Charcoal on paper, $19\frac{5}{8} \times 25\frac{1}{2}$ in. (49.8 × 64.8 cm.). New York, Metropolitan Museum of Art, the Alfred Stieglitz Collection.

152 (*above*) Ardengo Soffici, *Decomposition of the Planes of a Lamp*, 1912. Oil on board, $13\frac{3}{4} \times 11\frac{3}{4}$ in. (34.9 × 29.8 cm.). London, private collection.

153 Charles Sheeler, *Upper Deck*, 1929. Oil on canvas, 29⅛ × 22⅛ in. (74 × 56.3 cm.). Cambridge, Mass., Fogg Art Museum, Louise E. Bettens Fund.
154 Gerald Murphy, *Razor*, c. 1922. Oil on canvas, 32⅝ × 36½ in. (82.9 × 92.7 cm) Dallas Museum of Fine Arts, Foundation for the Arts Collection, Gift of Mr. Gerald Murphy.

Coalbrookdale a century earlier. But they did identify themselves with industry and its consequences far more unequivocally than any of their predecessors. Nor were they completely isolated. At the time when the first Futurist pictures were being painted, the Frenchman Robert Delaunay was busy turning the Eiffel Tower rather belatedly into a contemporary icon, celebrating it in his own quasi-Futurist work though it had been built as long ago as 1889.

The Futurists' paintings of crowds and machines in motion were perhaps their most spectacular achievements, but they did tackle other subjects as well. They even made Futurist versions of traditional still-life. Ardengo Soffici's *Decomposition of the Planes of a Lamp* (fig. 152) takes as its principal motif a banal mass-produced object. Soffici treated it in a way which gave it a new and startling authority. The Cubists, too, gloried in the banality of much of their

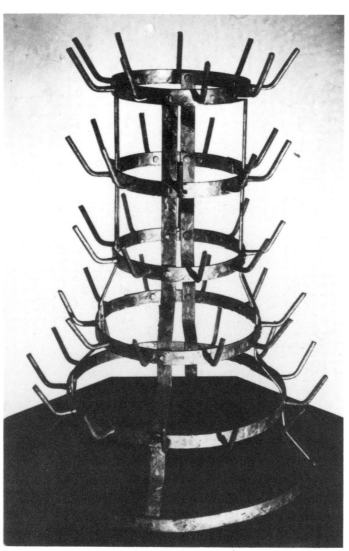

155 (*left*) Marcel Duchamp, *Bottle Rack*, 1914. Photograph by Man Ray, 1923. Arturo Schwarz Archives, Milan.

source material. The *collage*—the key invention of Synthetic Cubism—featured scraps of newspaper, old labels, fragments of wallpaper, in fact all kinds of industrial detritus. The invented 'reality' of art was brought into shocking juxtaposition with the kind of reality that surrounded everyone. The Dadaists, particularly Duchamp, took matters even further, presenting mass-produced objects completely unaltered within a fine art context (fig. 155). The ironic suggestion was made that these be looked at not as objects of use but as formal inventions.

Three things established themselves at the very heart of the modernist aesthetic, and continued to influence artists long after Futurism had exhausted its impetus. One was the cult of the machine itself.

156 (*above*) Fernand Léger, *Eléments mécaniques*, 1922. Drawing, $10\frac{1}{4} \times 8$ in. (26 × 20 cm.). Brussels, Musées Royaux des Beaux Arts de Belgique.

157 Fortunato Depero, *Steel and Turbine*, 1934. Rovereto, Galleria e Museo Depero.

Machines could be treated in a number of different ways—as a basis for abstraction, as in the impressive drawings of *Mechanical Elements* which Fernand Léger did in the early 1920s (fig. 156); or straightforwardly and for their own sake, but with a limpid clarity of vision which allowed the forms to make their own point. This was the method adopted by Charles Sheeler and other American Purists (fig. 153). Or, finally, machines could be allegorized, as in the mechanistic compositions painted by the Italian Fortunato Depero (fig. 157), who had once been an associate of the Futurists.

The second development was perhaps subtler, and also further-reaching in its effects. Duchamp presented ordinary mass-produced objects as if they were works of art. Other artists, less radical than he, took them into their vocabularies as subjects for painterly transformation. The American artist Stuart Davis, heavily influenced by French Cubism, took the Lucky Strike package as the subject-matter for a picture (fig. 158). Even before Raymond Loewy redesigned it, this package was one of the most familiar and ordinary of twentieth-century American objects. Davis asked his audience to shift focus and look at it in a totally different way, almost as if they had never seen it before.

Another American painter, Gerald Murphy, already seems to anticipate the Pop Art of the 1960s in a canvas produced in 1922. A matchbox, a safety-razor and a fountain pen are presented in quasi-heraldic fashion, almost as if they were images on an inn sign (fig. 154). Murphy seems to be saying that these industrial products, trivial and little considered, are in fact the emblems of a whole civilization and tell more about it than things with much greater pretensions to significance.

The fascination with machine forms had an

158 Stuart Davis, *Lucky Strike*, 1921. Oil on canvas, 33¼ × 18 in. (84.5 × 45.7 cm.). New York, Museum of Modern Art, gift of the American Tobacco Company, Inc.

inevitable impact on the decorative arts. Luxury products acquired an added frisson when they imitated what factories produced by the thousands or even the millions. Parisian jewellers made pendants in the shape of shells for heavy guns (fig. 160), and bracelets that seemed to be studded with ball-bearings (fig. 161). These fashionable follies were nevertheless a symptom of something important. People had started to study the products of industry in a new way, to savour industrial logic for its own sake. It is not too much to say that modern art, by separating industrial forms from their context, and holding them up to be admired in isolation, robbed industry of its innocence.

There was, of course, a continuing dialogue between the fine and the decorative arts, though this was seldom carried through as thoroughly as in Rietveld's famous chair (fig. 159), expressing the ideas

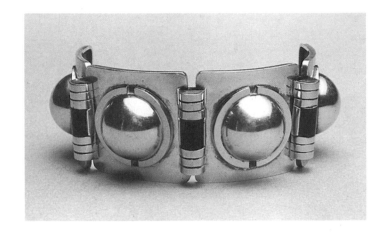

159 (*left*) Gerrit Rietveld, Red and Blue Chair, 1918.

160 (*right*) Jean Fouquet, shell pendant, *c.* 1925. Private collection.

161 (*above right*) Jean Desprès, Bracelet, *c.* 1930. Paris, Musée des Arts Décoratifs.

of the Dutch De Stijl group. French clocks, which a century earlier had been made to look like Greek temples in miniature, were now disguised as pieces of Cubist sculpture (fig. 162). A Swedish sculptor, Gösta Adrian-Nilsson, made a kind of Cubist idol with a loudspeaker in its head (fig. 163)—something not fundamentally different in intention from the most famous of all pieces of nineteenth-century kitsch—the porcelain statuette of the Venus de Milo with a clock-dial inserted in her stomach.

But there was a different kind of dialogue as well. In the nineteenth century, it had sometimes seemed as if pure machine forms were invisible. They only acquired visibility once they were ornamented in some way. Now art had endowed them with a kind of moral authority of their own. Design ceased to be pragmatic; men began to think of industry not as a brute force barely under the control of those who had created it, but as the paradigm of an ideal world. The machine must now be allowed to suggest its own forms and images, rather than having these imposed upon it by ignorant mankind.

162 (*far left*) Jean Goulden, Cubist clock, 1929. Private collection.

163 (*left*) Gösta Adrian-Nilsson, loudspeaker, *c.* 1920. Wood and metal, painted, height 22 in. (56 cm.). Sockholm, Moderna Museet.

164 (*right*) Karl Richard Henker, Electric table-lamps, illustrated in the yearbook of the Deutsche Werkbund, 1912.

The German Werkbund

If the Modern Movement in art prepared the way for new attitudes towards design in the broadest sense, in a much narrower sense the change was due to a group of artists, architects, craftsmen, manufacturers, bureaucrats and politicians in Germany. In October 1907 these banded themselves together to form the Deutscher Werkbund, a sort of design pressure group that held annual meetings in different German cities and formed regional Werkbund associations throughout Germany. Its declared aim was 'to ennoble industrial labour through the co-operation of art, industry and handicraft, by means of education, propaganda and united action on relevant questions'.

The single individual with the greatest responsibility for the Werkbund idea was a civil servant in the Prussian Ministry of Trade, Hermann Muthesius. Muthesius had trained as an architect, and in 1896 he was appointed architectural attaché at the German Embassy in London. He remained there until 1903, sending regular reports on English architecture, the progress of the crafts and industrial design. In addition he gathered information for a massive three-volume work, *The English House*, published in 1904–5, shortly after Muthesius's return to Germany.

Like many foreigners, before and since, Muthesius was struck by English pragmatism, and especially by the way in which this affected domestic arrangements. 'The genuinely and decisively valuable feature of the English house', he wrote, 'is its absolute practicality.' He was versed in the theories of Ruskin and Morris, but his special enthusiasm was reserved for the 'free' architecture of Norman Shaw—free because it broke away from formal, centralized planning. He was also keenly interested in the Garden City movement, which seemed to transfer this new attitude towards planning from buildings to the creation of whole communities. He was, on a slightly different tack, a friend and admirer of Mackintosh, and instrumental in spreading the latter's reputation in Central Europe.

Muthesius was not the only German enthusiast for the English Arts and Crafts Movement at this time. At the end of the nineteenth century, Germany had developed its own, rather belated, version of Art Nouveau, the Jugendstil, named after the Munich periodical *Jugend*, whose first number was published in January 1896. Jugendstil rather cautiously mingled French, Belgian and British influences—one early channel for English ideas was the patronage offered by the young Grand Duke Ernst Ludwig of Hesse, son of Queen Victoria's favourite child, Princess Alice. In 1897-8 the English architect Hugh Baillie Scott, aided by Ashbee and the Guild of Handicraft, carried out a major decorative scheme for the Grand Duke's new palace at Darmstadt.

In creating the Werkbund, Muthesius had two associates who were more prominent than the rest. One was the Belgian architect Henri van de Velde, who had settled in Germany in 1899, and who in 1906 had become the head of the Weimar School of Applied Arts—the predecessor of the Bauhaus. Van de Velde was a socialist much influenced by the political ideas of Morris. Though working in a recognizably Jugendstil idiom, he rejected the notion of 'art for art's sake', and believed in the necessity of a reconciliation with the machine.

The other partner was a former Protestant pastor and now prominent liberal politician called Friedrich Naumann. Indeed, it was Naumann's very necessary diplomatic skills which brought the founding members together. His involvement was due to the fact that he had always been an enthusiast for the visual arts, and, more specifically, an advocate of the need to find new forms to suit the modern age. At this period he was obsessed with the idea of bringing about a revival of German culture and of making it something which benefited rather than suffered from the machine.

Though the Werkbund survived the First World War and lived on throughout the whole of the turbulent Weimar years, to collapse only when the Nazis came to power, it was too large ever to be really united or homogeneous. Its membership rose from 492 in 1908 to a peak around 3,000 in 1929. To exist on this scale it had to be a forum of debate, rather than a tightly-knit pressure group pursuing well-defined ends. Its ranks were split by frequent rows.

One of the most virulent of these occurred in July 1914, on the eve of the Werkbund's Cologne exhibition —the most ambitious of all its pre-war manifestations. It was triggered off by a series of ten theses proposed by Muthesius for adoption by the Werkbund Congress. Muthesius wanted designers to concentrate henceforth on the development of standard or typical forms—things which could be manufactured in quantity to meet the needs of the export trade. Muthesius was opposed on the one hand by up-and-coming architects like Walter Gropius and Bruno Taut, who saw in his propositions an attempt to give the status quo the force of law; and on the other hand by van de Velde and others, who still valued

Jugendstil individualism. For the latter what Muthesius was advocating was a whip-hand for the manufacturer, while the designer would once more be reduced to the status of a mere pattern draftsman. Many of Muthesius's opponents were even hostile to the idea that Germany should try to export on a large scale. For them this meant a dilution of German folk identity, while at the same time pandering to debased foreign tastes. It also meant the sacrifice of quality in favour of cheapness. Muthesius was forced to back down in order to keep the Werkbund together.

This debate within the Werkbund on the eve of the war showed how far some members at least had moved from the English Arts and Crafts ideals which had originally inspired its foundation. In fact, Muthesius had never accepted English Arts and Crafts philosophy uncritically. His aim was to see what could be learned from Britain for Germany's benefit. It was a period of ever-increasing rivalry between the

two countries, the most conspicuous symptom being the naval arms race. Educated German opinion swung between admiration of England and dislike motivated by envy. Germany had achieved unity only in the nineteenth century. Industrialization had taken place much later than it did in England, and Germans were conscious of having a great deal of ground still to make up. On the other hand, because German industry was newer, it was less hidebound, more willing to try new methods.

This did not mean blanket approval for the capricious experimentalism of Art Nouveau, which had already manifested itself in a number of designs which still look avant-garde today. Richard Riemerschmid, who played a prominent part in the early councils of the Werkbund, was previously connected with the Deutsche Werkstätte—equivalents of the craft workshops which had been created in Vienna. In fact they had come into being six years earlier than

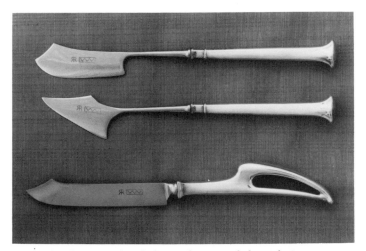

165 Richard Riemerschmid, dinner, butter and cheese knives, *c.* 1900, made for the Deutsche Werkstätte.

166 Henri van de Velde, cutlery, illustrated in the yearbook of the Deutsche Werkbund, 1912.

167 Henri van de Velde, Tennis Club, Chemnitz, illustrated in the yearbook of the Deutsche Werkbund, 1912.

168 Walter Gropius, desk and chair, illustrated in the yearbook of the Deutsche Werkbund, 1912.

those in Vienna, in 1897. Riemerschmid's cutlery, designed for these workshops in 1900 (fig. 165), marks a startling breach with conventional forms, an attempt to rethink table-knives and the way we use them from first principles. For all their undoubted elegance—a quality also to be seen in Riemerschmid's furniture designs of the same period—these knives must have seemed extremely outré then, and they are still so today. Henri van de Velde's cutlery, exhibited in the Werkbund exhibition of 1912, is a very different matter (fig. 166). It shows subtle modifications, and above all simplifications, of already accepted shapes. One sees exactly the same principles at work in van de Velde's understated interior for a tennis club in Chemnitz (fig. 167). Even more revealing in its way is the furniture by Gropius featured in the Werkbund Yearbook for 1912

(fig. 168). The massive desk and matching chair show an obvious reversion to the Biedermeier style, which was an unacknowledged inspiration of much 'progressive' design at this period, especially design which was now reacting against Art Nouveau. Though both van de Velde and Gropius featured so prominently among Muthesius's opponents at the Werkbund meeting of 1914, both of them here seem to come close in practice to the rules he tried to lay down.

The most seminal Werkbund designs were not for things such as these, which had existed in one form or another for centuries, but for things which were a response to technological change, and most of all to the increasing use of electricity in the home. Richard Schulz's pendent light-fitting is a modification of designs already evolved for gas, but the cleanness of line is something new (fig. 169). A series of table-lamps by one of the less-known Werkbund designers, Karl Richard Henker of Charlottenburg (fig. 164), shows a logic in approaching this particular problem which is rare even today.

It is no accident that the most frequently cited domestic designs of the period are those of Peter Behrens, also a prominent Werkbund figure. Behrens, a painter turned architect, is often called 'the first industrial designer'. That honour belongs more justly to Christopher Dresser. Behrens was something different—the first house-designer, responsible for the visual impact made by a large industrial corporation.

Behrens joined AEG in 1907, the year the Werkbund was founded. The Allgemeine Elektrizitäts-Gesellschaft had been created by a Jewish engineer called Emil Rathenau, who acquired German rights to Edison's patents for electric lighting systems. By 1907 it was one of the biggest manufacturers in the world, producing generators, cables, transformers, motors, light bulbs and arc lamps. The founder's son, also very much part of the firm, was Walther Rathenau, later to be Foreign Minister under the Weimar Republic. It was Walther Rathenau who described AEG, at this precise moment, as 'undoubtedly the largest European combination of industrial units under a centralized control and with a centralized organization'. This centralization was of course crucial to Behrens's role. After his first work for the company (designs for arc lights) was pronounced a success, Behrens went on to design not only numerous industrial products, but the company's factory buildings and the advertisements and printed material it put out. Everything showed the impress of a single creative mind.

Behrens was in some ways rather a conventional designer—much of his domestic architecture is a bland version of Neoclassical style; his overbearing German Embassy in St Petersburg was a favourite building of Adolf Hitler, and provided Speer with a model for the colossal structures planned to celebrate the achievements of the Third Reich. On a far smaller scale, some of Behrens's designs for household objects would set the teeth of design historians on edge were they not by him. A case in point is the electric kettle illustrated in a previous chapter (fig. 56).

However, Behrens was not afraid of the vast opportunity which had been offered him, and much of his design thinking benefited from a firm grasp of classical forms. He built up a good collection of Greek vases, and it is clear that he often consulted these for inspiration. A humidifier made to project from a ceiling (fig. 170) is a kind of Greek *kylix* turned upside down. The resemblance is so strong that it is nowadays sometimes reproduced the wrong way up.

169 Richard L. F. Schulz, pendent lamp, illustrated in the yearbook of the Deutsche Werkbund, 1912.

170 Peter Behrens, humidifier, 1909, manufactured by AEG.

Most important of all, in terms of the future, are designs where Behrens was content to allow a machine to make its own statement, even in a domestic environment. There is nothing disguised or apologetic about the table-fan he designed around 1908 (fig. 171), and no attempt is made to force it to conform to its presumed surroundings. Yet even here the architect's instinct for form and balance is in evidence—most conspicuously in the square block at the top of the conical stem, just below the yoke which supports the mechanism. Behrens may have taken some hints here from traditional designs for scientific instruments, perhaps prompted by the fact that the angle of the fan itself needed to be easily and conveniently adjustable—the kind of problem which makers of telescopes had been solving for several centuries.

171 Peter Behrens, electric fan, *c.* 1908, manufactured by AEG.
172 Le Corbusier, 'Basculant' chair, chrome steel tube and leather, 1929.

The Triumph of Modern Design

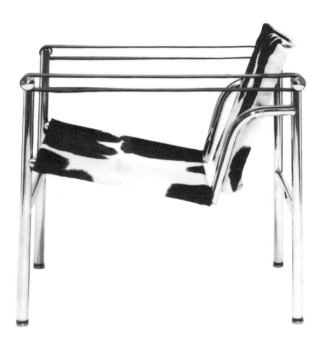

It is at least arguable that the real credit for firmly marrying design to the Modern Movement belongs not to the German artists and architects who transformed and extended Werkbund principles in the post-war years, using the Bauhaus as their laboratory, but to members of the Dutch avant-garde art-group, De Stijl. De Stijl took its name from the periodical of the same name, edited by Theo van Doesburg from 1917 to 1931. It was not a group in the usual sense, in that many of the chief participants knew each other only slightly, while two of the most important—Gerrit Rietveld and Piet Mondrian— never in fact met at all. But there was nevertheless a real ferment of ideas in Holland, especially during the period 1917 to 1921 when the country was isolated by the war and its aftermath. The puritan simplicity of traditional Dutch Calvinism was married to radical new ideas about the visual arts. Painting and sculpture were not considered to be separate from architecture or furniture design—they were all part of the same statement. The nature of this statement can be deduced as easily from Gerrit Rietveld's famous wooden armchair, designed in 1917 or 1918 (fig. 159), as it can from a painting by Mondrian. Rietveld said of it: 'The construction is attuned to the parts to insure that no part dominates or is subordinate to the others.

In this way the whole stands freely and clearly in space, and the form stands out from the material.' It must also be added that the Rietveld chair is not at all comfortable to sit in. It is a sculpture in the shape of a chair, rather than a practical object.

More functional, and even more prophetic for the future, is the hanging light Rietveld designed a few years later (fig. 173). Here, ordinary commercial tube lights are held by small black-painted blocks and are then suspended—two horizontally, one vertically. The result is a practical, totally unpretentious light-fitting which is also an abstract sculpture in space— the ancestor of the abstract sculptures made of neon tubes which were the *dernier cri* in the late 60s and early 70s.

The Bauhaus, founded by Walter Gropius in 1919 as a result of the amalgamation of the two main schools of art in Weimar, including the one which van de Velde had headed before the war, started by pursuing what were apparently very different ideals to those professed by the artists of De Stijl. Gropius said, at the time he founded the Bauhaus, that his aim was to create 'a happy working community such as had existed in an ideal way in the masons' lodges of the Middle Ages.' The Bauhaus Founding Manifesto, signed by Gropius, declared:

> Architects, sculptors, painters, we must all turn to the crafts. Art is not a 'profession'. There is no essential difference between the artist and the craftsman. The artist is an exalted craftsman. In rare moments of inspiration, moments beyond the control of his will, the grace of heaven may cause his work to blossom into art. But proficiency in his craft is essential to every artist. Therein lies a source of creative imagination.
>
> Let us create a new guild of craftsmen, without the class distinctions which raise an arrogant barrier between craftsman and artist. Together let us conceive and create the new building of the future, which will embrace architecture and sculpture and painting in one unity and which will rise one day towards heaven from the hands of a million workers like the crystal symbol of a new faith.

True enough, Gropius was later to imply that all this was largely a smokescreen—a matter of getting the whole project going in the adverse conditions of the time. If this was truly the case, his choice of man to run the Bauhaus Preliminary Course, the indispensable foundation of all the rest, was rather strange. The job went to the mystical Johannes Itten, who based his teaching on total spontaneity and freedom of expression. He even used to start his sessions with breathing and vibration exercises. It was only in the spring of 1923, when Gropius announced that he could no longer defend Itten's methods, and the latter departed, that a different and more practical regime began.

The Bauhaus was unique for its time in combining distinguished fine artists—among them established members of the avant-garde such as Klee and Kandinsky—with a school of design. But where design itself was concerned the institution's watch-word was 'functionalism'. Like all words which are taken up and turned into slogans, this one acquired a special nuance. Functionalism meant obedience to the demands made by Theo van Doesburg, who came to Weimar in person in January 1921, and paid shorter and longer visits thereafter. 'Proportion was to take the place of form, synthesis to replace analysis, logical construction lyrical suspension, mechanics craftsman-ship, collectivism individualism; such were the demands of De Stijl which in architecture followed the principles of order of Mondrian' (Ludwig Grote, catalogue of the exhibition *50 Years Bauhaus* held at the Royal Academy in London in October 1968).

Another important influence was the Swiss-born

173 Gerrit Rietveld, hanging light, 1921.

architect Le Corbusier, who had worked pre-war in Behrens's office, where he got to know both Gropius and Gropius's colleague Ludwig Mies van der Rohe. In October 1918 Le Corbusier, now settled in Paris, published the first number of a magazine called *L'Esprit Nouveau*. His colleagues in the venture were the painter Amédée Ozenfant and the poet Paul Dermée. Le Corbusier used the periodical as a vehicle for his ideas about architecture—he preached the superiority of simple geometric forms—'their image appears to us pure, tangible and clear-cut'. He also repeatedly put forward the idea that a house is 'a machine for living in', by which he meant that it must serve contemporary needs without compromise, and must function with the precision of a machine.

The Bauhaus rapidly absorbed these doctrines, and distinctive designs began to emerge from its studios. Their commercial success, though limited to an élite market, helped to keep the institution itself alive. Some of these designs had definite affinities with the vernacular industrial products of the nineteenth century. Theodor Bogler's Mocca machine—one of the best-known Bauhaus designs, manufactured by the State Porcelain Factory in Berlin (fig. 174)—resembles Wedgwood Queen's Ware kitchenwares and their Biedermeier equivalents (fig. 50). Even simpler and more authoritative in shape is Wilhelm Wagenfeld's tea-service of heat-resistant glass (fig. 175). Forms of this sort gained a wide currency by the beginning of the 30s, and in some Bauhaus influence can be seen operating indirectly rather than directly. Hermann Gretsch's inexpensive and hugely successful design 'Arzberg 1382' (fig. 176), winner of gold medals at the Milan Triennale of 1936 and the Paris World Fair of 1937, is a more graceful version of a design by Bogler done as early as 1923.

In some Bauhaus designs it is possible to see the

174 (*left*) Theodor Bogler, mocca machine for production in series, 1923.

175 (*below far left*) Wilhelm Wagenfeld, tea-service in heat-resistant glass, 1933.

176 (*below left*) Hermann Gretsch, 'Arzberg 1382' tea-service, 1931.

177 (*right*) Wilhelm Wagenfeld, table-lamp, 1924. Munich, Neue Sammlung.

178 (*below right*) Marcel Breuer, cantilever chair, tubular steel, bentwood and cane, 1928.

withstand severe static strain (tractive stress of the material). The light shape increases the flexibility. All types are constructed from the same standardized, elementary parts which may be taken apart or exchanged at any time.

This metal furniture is intended to be nothing but a necessary apparatus for contemporary life.

Oddly enough Bauhaus furniture, and other metal furniture affected by the Bauhaus idiom, now seems to have a very definite period character of its own. When made under today's conditions, the designs that have survived cannot be sold at mass-market prices—they are prestige objects. This is perhaps forgivable in the case of Mies's Barcelona chair and stool (figs. 180 and 181), which were originally a kind of 'palace furniture'

influence of the creations of the pre-war Werkbund. This is true of Wagenfeld's well-known table-lamp of 1924 (fig. 177), which is a rehandling of a series of designs published in the Werkbund Yearbook for 1912 (fig. 164).

The thing most people now associate with the Bauhaus, however, is furniture made of metal, and in particular of metal tubing used on the cantilever principle. There is some dispute about who first had the idea—it may in fact have come from the Dutch architect Mart Stam, who was not a member of the Bauhaus at all. Breuer (fig. 178) and Mies van der Rohe also did versions, and Breuer was also responsible for an extremely ingenious easy chair of more ample proportions (fig. 179). In Breuer's eyes, furniture of this type was styleless. Writing in 1928, he explained:

I purposely chose metal for this furniture in order to achieve the characteristics of modern space elements . . . the heavy imposing stuffing of a comfortable chair has been replaced by a tightly fitted fabric and some light, springy pipe-brackets. The steel used, and particularly the aluminium, are remarkably light, though they

used to articulate the spaces in the German Pavilion at the Barcelona World Fair of 1930. It is less so in that of Le Corbusier's Basculant armchair of 1928 (fig. 172), especially as Le Corbusier himself was an impassioned advocate of 'off-the-peg' furniture of the cheapest possible sort. His revolutionary Pavillon de L'Esprit Nouveau, a counterblast to the exponents of Art Deco in the Paris Exposition des Arts Décoratifs of 1925 (the event from which Art Deco took its name), featured furniture of this type, and particularly chairs from Thonet (fig. 182).

Perhaps because of their interest in Thonet, Bauhaus designers and those who fell under their spell were also interested in the possibilities offered by steam-bent plywood. Breuer, forced into exile by the coming of the Nazis, produced designs utilizing this material for the English firm of Isokon (fig. 183). He did not, however, manage to take it as far as the British designer Gerald Summers, whose armchair, manufactured by the appropriately named Simple Furniture Co., *c.* 1934, is cut and bent from a single plywood sheet (fig. 184). Here virtuoso use of the material

clearly outran the practical demands of comfort.

It was, however, metal not wood which took the leading place, because metal was the symbol of the machine age. Even the veteran Frank Lloyd Wright could not resist exploring its possibilities, and did so with characteristic originality in an office chair designed in the mid-30s for Johnson Wax (fig. 185). In Paris, designers working for a rich and fashionable clientele, who a few years previously would have demanded exotic and expensive hardwoods, now turned to chromed metal. The Parisian designer René Herbst made a sumptuous dressing-table in chromed and painted metal and mirror-glass for the Princess Aga Khan (fig. 186), and Adrienne Gorska designed a scarcely less sumptuous one for her sister the painter Tamara de Lampicka (fig. 187).

These commissions were a sign that the new aesthetic had penetrated a certain kind of consciousness. In Germany, before the Stock Market Crash of 1929, which crippled the still shaky German economy, there were even signs that the Bauhaus message was starting to carry beyond the intellectuals and the rich, and to

179 (*far left*) Marcel Breuer, Wassily chair, tubular steel and leather, 1925. London, Victoria and Albert Museum.

180, 181 (*left*) Mies van der Rohe, Barcelona chair and stool, welded steel strip with rubber strips and leather upholstery, 1929.

182 (*above*) Le Corbusier, living-room in the Pavillon de l'Esprit Moderne, at the 1925 Paris Exposition des Arts Décoratifs.

183 Marcel Breuer, bentwood chaise-longue for Isokon, 1936. London, Victoria and Albert Museum.

184 Gerald Summers, bentwood chair for Simple Furniture Ltd. *c*. 1934. London, Victoria and Albert Museum.

185 Frank Lloyd Wright, office chair for the Johnson Wax building, in painted steel and walnut, 1936–9.

install itself in the general bourgeois consciousness. There were none that it was making any substantial impact on the mass market, despite Bauhaus concern for topics such as working-class housing.

Strangely enough, industrial design managed to consolidate its status where industry itself was concerned only in a very different environment, and in a very different set of circumstances.

The small group of pioneering industrial designers in America—among them Raymond Loewy, Norman Bel Geddes and Walter Dorwin Teague—came from a different background from the intellectuals who founded the Bauhaus. Loewy emigrated from France to the United States, and his background included a

few months working for an electrical engineering firm. His first career was as a fashion illustrator in New York, which eventually broadened to include designing advertisements as well as simply doing the drawings for them. His first job as a designer was a cosmetic one—he tidied up the Gestetner duplicating machine (fig. 188). It was some time before he managed to get another of equivalent importance, till the Depression suddenly convinced manufacturers that 'styling', as it was called, might be the remedy.

Loewy's chief rivals came from very different backgrounds and possessed equally contrasting temperaments. Teague was born in 1883, and was the son of a small-town Methodist minister. He went into

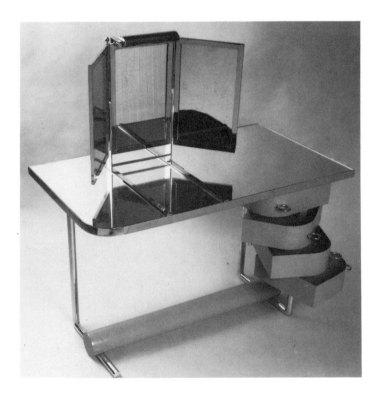

186 (*below far left*) René Herbst, dressing-table designed for Princess Aga Khan, 1930.

187 (*below left*) Adrienne Gorska, dressing-table designed for Tamara de Lampicka.

188 Raymond Loewy, Gestetner's '160 Express Inker', redesigned 1929.

and he, too, started his career as an advertising man. From this he graduated to stage design, at which he was very successful—his greatest triumph in this field was the New York production of Reinhardt's *The Miracle* in 1923, for which he transformed the interior of the Century Theatre into a convincing replica of a Gothic cathedral. He then became a highly successful designer of extremely theatrical shop window displays, and graduated via these, thanks to a connection with the big advertising agency J. Walter Thompson (which was then run by an uncle of his second wife), to designing or redesigning utilitarian objects. Bel Geddes's schemes were accepted less frequently than those of his rivals, and he had something of a reputation for losing his patrons money. But his flamboyant book *Horizons*, published in 1932, made him the leading popular 'futurologist' of his period, and his reputation in this sphere culminated in an invitation to create a Futurama depicting the world of 1960 which was the main feature of the General Motors pavilion at the new York World's Fair of 1939. It was an immense hit with visitors to the Fair.

Both Loewy and Bel Geddes were leading exponents of what came to be called the 'Streamline Style', as sleek and curvilinear as Bauhaus design was four-square. Loewy applied it appropriately to locomotives (fig. 191) as well as to long-distance buses and automobiles. One of Bel Geddes's most ambitious projects was for a totally streamlined ocean liner, with all its decks enclosed. His model for this dates from 1932 (fig. 190), and still looks seductively novel today. It was not as much in advance of its time as one might now imagine, since a large ferry of similar design, the *Kala-Kala*, was in operation between Bemerton and Seattle on Puget Sound in the mid-1930s (fig. 189). Streamlining, however, became so popular that it was

advertising and made a reputation for dignified selling of quality goods. From a study of seventeenth- and eighteenth-century French history and culture he graduated to a fascination with what was going on in Europe in his own time, and in 1926 voyaged to Europe to study the work of Le Corbusier and Gropius among others. But he also looked carefully at the more conservative French decorators who were creating the Art Deco style for rich private clients. And on his return he applied what he had learned to strictly commercial ends, such as interiors for high-class shops. From this it was a short step to the design of consumer goods, such as automobiles. Teague's aim was always to increase profits for his clients, but without too much sacrifice of aesthetic integrity. His designs used existing production techniques, and improved familiar forms through omission and simplification.

Norman Bel Geddes also came from a small town,

189 The streamlined ferry-boat *Kala-Kala*, in operation on Puget Sound in the 1930s.

190 Norman Bel Geddes, project for a streamlined ocean liner, 1932. University of Texas at Austin, Hoblitzelle Theatre Arts Collection, Humanities Research Centre.

191 Raymond Loewy, T-1 locomotive for the Pennsylvania Railroad, *c.* 1946.

applied to anything and everything, even the most apparently static of domestic appliances (fig. 192), though the electric toaster illustrated is stylistically provincial—the product of a British firm aping American modes. Objects of this kind do help to explain why the phrase 'industrial design' acquired a negative connotation for many American intellectuals. In his autobiography, Buckminster Fuller described it roundly as 'the greatest betrayal of mass communication integrity in our era'.

Yet there are also reasons for thinking that the developments in America in the early 30s tell us more about the true nature of industrial design than anything which took place at the Bauhaus. American design at last made the breakout from the domestic sphere. And it became directly linked to marketing— what people were supposed to want, or ought to want, yielded to a consideration of what they actually did want. This, in turn, brought a further step in the chain of reasoning—*why* did people want what they appeared to want? Once consumer motivation was firmly linked to a rapidly advancing technology, industrial design acquired a new and extremely functional role—it confronted what was already familiar to the consumer with what science and engineering were making possible.

192 Salmic streamlined electric toaster, in current production 1947.

Designers Today

The success of the leading American designers of the 30s set the tone for the next two decades, far more so than did the theories of their Bauhaus colleagues and predecessors. Nevertheless, thanks to political events, a certain amalgamation of European and American ideas took place. The leading members of the Bauhaus were forced out of Germany by the rise of Nazism, and America was the place where the majority of the exiles tried to rebuild their careers. They were thus brought into direct contact with the American industrial scene. But their careers in America were more, rather than less, oriented towards architecture, as it was here that they found opportunities to build which the Weimar Republic had been unable to provide. For example, the majority of Mies van der Rohe's buildings went up on American soil, though they were often based on studies made in Germany. Preoccupied with building, the men of the Bauhaus tended to move away from the industrial sphere, which had always been a kind of substitute arena. Mies van der Rohe built a vast campus for the Illinois Institute of Technology, and skyscraper apartment blocks on Chicago's Lakeshore Drive, but it was Loewy (with no background in either architecture or engineering) who remained in close touch with the Pennsylvania Railroad and with Studebaker.

It was very largely through the efforts of men like Loewy that industrial design established itself as a viable career. Loewy liked publicity, and he was adept at organizing a luxurious life-style for himself. This counted for almost as much as the growth of the professional associations which, from the teens of the century, had followed in the wake of the Werkbund. Among these was the British Design and Industries Association, set up in 1915.

It was established that a designer could work with industry in various ways. Like Raymond Loewy, he could head, or at least belong to, an outside design studio. Such a studio could be called in on a one-off basis, to undertake a specific task, or could be retained on long-term contract as a consultant. Alternatively, a given firm could establish its own team of designers, though there was a wide choice as to how such a team fitted into the company structure. In general, if the operations of a particular company were allied to heavy engineering, the design group would be called in at an early stage and might be allied to the research and development department. Designing an automobile would mean re-tooling to a greater or lesser extent, so it was essential to call in the designer as soon as possible.

On the other hand, if the company's products were consumer perishables or disposables rather than consumer durables, the designer was usually only consulted at a much later point. His work was felt to be akin to advertising, and the design unit, concerned chiefly with packaging, might well form part of the advertising department.

The idea that design and engineering could be thought of as being one and the same thing took a long time to establish itself—something which Loewy's career, in particular, can be used to demonstrate, since so many of the tasks assigned to him turn out to have been a cosmeticization of products whose basic form had already been determined by professional engineers.

One subtle effect of the industrial depression of the 30s was to drive men into industrial design who might otherwise have had viable careers as fine artists. This switch in direction was also attuned to the aesthetic mood of the decade, which put great stress on the idea of the 'useful' arts, as opposed to visual art which existed without apology, for its own sake. An example of such a change in career is provided by the gifted young English print-maker William Larkins—a contemporary of Graham Sutherland. Like Sutherland, Larkins was badly affected by the collapse of the

194 (*left*) Charles Eames, glass fibre side chair, designed for Herman Miller, 1949.

195 (*right*) Eero Saarinen, 'Tulip' dining chair, designed for Form International, 1956.

193 (*opposite*) William Larkins, 'Black Magic' chocolate box, designed for Rowntree & Co., 1934.

196 William Larkins, *Bush House*, 1927. Etching and engraving, 8 × 9½ in. (20.3 × 24.1 cm).

197 (*opposite*) Wallis, Gibson and Partners, Hoover factory, Perivale, Middlesex, 1932–5.

boom in modern prints which took place at the end of the 20s. He made a second career in advertising and design, and is now remembered as much for the Black Magic chocolate box (fig. 193) as he is for his masterly print of Bush House in London (fig. 196).

From the 30s to the 50s industrial design enjoyed increasing prestige with the public. Considerable propagandist work was done through design exhibitions, and official organizations were created to help the cause of good design. In Britain, the Council of Industrial Design was founded in 1944, and the 'Britain Can Make It' exhibition was organized in 1946. The themes of this exhibition were restated on a more ambitious scale at the Festival of Britain of 1951. The then Director of the Council of Industrial Design, Gordon Russell, wrote in his foreword to the booklet, *Design in the Festival*, that design itself was to be 'recognized as an integral part of quality, which can no longer be thought of as good workmanship and good material only. In fact good design should be recognized as one of the consumer's guarantees of quality, as the firm which takes the trouble to design an article with real care and skill will certainly see that it is honestly made.'

A body of design literature started to be built up. In the Ruskinian tradition much of this was fiercely critical of industry and its relationship to design. One of the most important design critics in Britain was the German exile Nikolaus Pevsner, whose *An Enquiry into Industrial Art in Britain* was published in 1937. Much of this was devoted to a denunciation of the state of design at that moment. 'When I say that 90 per cent of British industrial art is devoid of any aesthetic merit,' Pevsner wrote, 'I am not exaggerating.'

Pevsner was committed to Bauhaus ideas and values, and was a fierce opponent of the Streamline Style, which he dismissed as 'bad modernism'. The Hoover Building in West London, now much admired by architectural historians as one of the best examples of popular Art Deco (fig. 197), Pevsner dismisses as being 'perhaps the most offensive of the modernistic atrocities along this road of typical by-pass factories'.

In fact, it is possible, from the 30s to the 50s, to see a relatively simple economic and aesthetic structure. Design purists were those most affected by Bauhaus ideas. The contemporary products they admired were usually, but not invariably, expensive, and found their market with educated upper-middle-class consumers. There were also products which expressed an excitement about the idea of being modern, a commitment to the future, without necessarily adhering to Bauhaus standards. Though the 'modernistic' style was rooted in the Art Deco of the Parisian 20s, it found its fullest expression in America, and persisted well into the 50s. Detroit automobile design of the 50s was still firmly married to this idiom. And finally, there was a large category which consisted of industrially made objects which did not seem to have been consciously designed at all.

Nevertheless, the idea grew up that design was something which could provide certain products with a competitive edge over their rivals. In the United States, far more than in Britain, consumers began to be conscious of major design reputations, and the work of 'name designers', such as Charles Eames or Eero Saarinen for furniture, enjoyed undoubted prestige on the market (figs. 194 and 195).

By the 50s it was established that there was a kind of aristocracy of design. Members of this aristocracy included big industrial corporations, such as IBM, which had become increasingly image-conscious, and manufacturers of the more expensive types of consumer goods. Below this level, design for mass consumption remained a free-for-all. The situation was disturbed by the advent of Pop Art in the 60s. Pop involved a kind of *trahison des clercs*—intellectuals now deliberately professed admiration for the cheapest and crudest elements in contemporary mass culture. This was followed by a modernistic revival in the 70s which involved a similar but less brutal rejection of what had hitherto been thought of as good design standards. In fact, the Bauhaus pioneers had been concerned to eliminate stylistic considerations altogether—they regarded style and logic in construction and manufacture as things inevitably opposed to one another. A new generation of post-war designers

got round this situation to their own satisfaction by putting greater and greater stress on the symbolic element in design, pointing out that manufactured objects must offer the consumer psychological as well as practical satisfactions and must cater to fantasy almost as much as they did to use. A new school of design criticism concentrated on this symbolic aspect and analysed (for example) the way in which manufacturers of portable wirelesses tended to take their cue from military communications equipment.

Nevertheless, the past two decades have also seen the growth of a more widespread design consciousness. This can be traced to the growth of art education in general. One effect of the Second World War was to legitimize the Modern Movement—largely because the Third Reich had persecuted it. This led not only to a spate of exhibitions of modern art to celebrate the cessation of hostilities, but to the expansion of the whole institutional apparatus of modernism: art schools and colleges as well as museums. Where the teaching of art and design was concerned, this more and more tended to fall into the hands of those who had some historical connection with the Bauhaus— people who had actually taught or studied there, or who were the disciples of the original Bauhausler. In particular, the Bauhaus Preliminary or Preparatory Course, a basic education in handling forms and colours, became standard in almost all schools teaching any course in art or design.

In addition, the Bauhaus had some notable direct successors—chief among them the New Bauhaus, founded in 1937, and later incorporated into the Illinois Institute of Technology; and the Hochschule für Gestaltung in Ulm, founded in 1955, and at first directed by Max Bill, who had belonged to the original Bauhaus. But the violent controversy between Bill and his deputy Tomàs Maldonaldo, which marked the

early years of the Hochschule, also served to demonstrate that it would be difficult to apply the old Bauhaus principles absolutely rigidly in a new and different era. Maldonaldo's point was that it was not enough simply to search for new forms, as the Bauhaus had done—designers must find the flexibility to fit in with the far more complex demands being made by post-war technology and industry. It was he who

198 Habitat furniture on display at the Habitat showroom, Old Fulham Road, London, 1964.

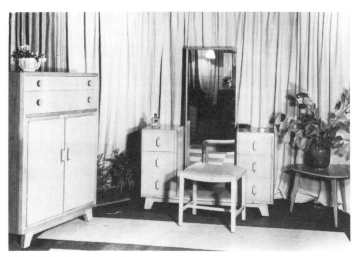

199 Utility furniture, 1948.

carried the day, and Bill who was forced to resign.

The art schools and colleges were in some ways more successful in affecting the general cultural atmosphere than they were in penetrating industry itself. In Britain, where industry remained extremely conservative in its attitudes towards designers and their use, it is probable that there was a considerable wastage of new talent, not least because the liaison between the teaching institutions and the world of practical work was not nearly as close as it should have been. On the other hand the art schools had a remarkable leavening effect on the community as a whole—it was no accident that so much of the new pop music was art-school connected—groups as different from one another as the Pink Floyd and Roxy Music sharing the same basic background.

The result was the gradual creation of a new

sensibility which had a profound effect on attitudes to design, and especially to design in the home. It is interesting in this respect to contrast the very different fate of wartime and post-wartime Utility furniture in Britain (fig. 199) and that of the goods sold by Terence Conran's chain of Habitat shops (fig. 198). Utility was an attempt to impose good taste on the mass public from above, using wartime and post-wartime shortages as the lever. The public rebelled as soon as it was able. Habitat offered a very similar kind of merchandise, though with much greater variety, and succeeded triumphantly. It was not merely that the gospel of good design had had time to make its impact, but that Habitat offered flexibility of choice within an overall stylistic unity—the kind of unity which had previously only been available in one-off interiors styled by professional designers.

Design for industry now affects so many aspects of daily living that it is hard to see the subject as a unity—a difficulty which affects not only writers about design but those who actually train and employ designers. The man who designs a motor car needs a different background from the man who oversees the production of a furniture factory, and both in turn are very different from the man whose professional concern is with a range of packaging. Yet all of these individuals are described as 'industrial designers'. The only two things one can see clearly are that design, for all its attempts to escape, remains inevitably linked to the consumer capitalism of the West, and that the puritanism and moralism which have been very much part of the design heritage, from the time of John Ruskin if not from that of Josiah Wedgwood, look increasingly inappropriate.

PART 2

CASE STUDIES

Designing the Automobile

As can be seen from the preceding chapters, there is a paradox about the history of industrial design. It is most easily told through domestic objects. These perhaps changed their shapes because of the Industrial Revolution, and in some cases the actual materials of which they were made. But they did not either change their functions or introduce entirely new functions which changed the nature of society itself. A drinking-glass remained a drinking-glass, whether or not it was industrially produced.

The non-domestic product which best demonstrates the nature of industrial design is the automobile. No other machine carries such complex

emotional overtones, and probably none has had quite such a major impact on people's everyday lives. In the development of the automobile we see a unique interplay between functional and symbolic design, and an examination of its history also brings home to us the fact that in designing an automobile there is no single satisfactory answer—successful design depends both on the aim the designer has in view, and on the social and technological factors which he has to take into account.

It is generally held that the automobile was invented by Carl Benz in 1885. It took the form of a tricycle powered by an internal-combustion engine, and it was followed the next year by Gottlieb Daimler's four-wheeler, which was a conversion of a horse-drawn cart (fig. 200). These were not, however, the first 'horseless carriages', as steam-powered versions had been in existence for some time. Technical improvements took place fairly rapidly after 1900. In 1901, the first Mercedes had a chassis frame of pressed steel; a honeycomb radiator; ignition by magneto; mechanically operated valves; and a 'scroll' clutch linked to a gate-change gearbox. In 1906, The French firm of Sizaire-Naudin introduced the idea of independent front suspension.

The technological development of the automobile, like technological development in many other fields, did not proceed smoothly and evenly but in a series of jerks. Nor were new ideas immediately and universally taken up throughout the industry. An innovation introduced by one marque might have to wait many years before it appeared elsewhere. Some innovations failed to make much immediate effect. The first

streamlined car was constructed in 1921 by Edward Rumpler (fig. 201). It also had all-independent suspension, and an engine at the rear, not the front. But these novel features made little impact; people presumably found the car too different from anything they knew. The First World War had established automotive transport as something absolutely universal, and there was already a norm for automobiles from which a designer departed at his peril. Other innovations were less visible, and perhaps therefore more successful in obtaining a foothold. In the inter-war years the huge resources of the American automobile industry provided the means for constant research, and refinements were continually being added in the hope of giving one make the edge against its rivals. This continued even during the Depression. The American firm of Packard pioneered a whole series of advances. In 1925 it introduced centralized chassis-lubrication; it made synchromesh gearboxes and vacuum servo-brakes standard on its products

200 (*left*) Gottlieb Daimler, with his son Adolf, driving their 4-wheeled automobile, 1886.

201 Streamlined saloon designed by Edward Rumpler, 1921. Munich, Deutsches Museum.

Illustration page 118. Packaging for the Meridian stereo system, 1981; cf. fig. 380.

from the early 30s; and Packard braking systems became hydraulic in 1937. It is significant that these innovations were aimed first at increased convenience for the user, and secondly at increased safety.

In Europe, too, certain manufacturers were technically very advanced. The 1928 Voisin 16.50 coupé (fig. 204), manufactured in France, had a panel body made entirely of pressed aluminium, which eliminated the old coachbuilding system of panels fixed to a wooden framework. There were anti-glare lamps and infinitely adjustable seats. There was also an unusual amount of glass for the time, though the shape of the body was relatively conventional for its epoch (cf. fig. 215). In this case, the combination of advanced specification and restrained looks clearly appealed to the well-heeled, and the Voisin attracted an impressive number of royal and show-business customers—among them Maurice Chevalier, Josephine Baker, the Queen of Yugoslavia and the King of Siam. Le Corbusier gave the marque a different and perhaps more significant accolade by buying one. Yet subsequently the Voisin failed to achieve classic status—that is, it failed to appeal to the imagination of car fanatics in the same way as the Bentleys, Hispano-Suizas and Bugattis of the same epoch.

The design history of the automobile was less influenced—and perhaps in a sidelong way the Voisin helps to make the point—by mechanical improvements than it was by the sociological framework within which this complex piece of engineering existed, and perhaps still more by the powerful fantasies it evoked. Taken simply as an idea, the automobile could be interpreted in a large number of different ways. At the start of its career, it was a rich man's plaything, and a little later it became a rich man's convenience. As it evolved further, it became possible to differentiate types, whose functions were

202 (*top left*) Bugatti Type 13, 1910.

203 (*centre left*) Standard 15 h.p. Landaulette, 1911.

204 (*bottom left*) Voisin 16.50 coupé, 1928.

205 (*right*) Production line for the Model-T Ford, Highland Park, 1913.

expressed, just as those of the old horse-drawn vehicles had been, by different styles of coachwork. Even luxury vehicles could cover a wide spectrum, depending on the tastes as well as the status of their owners. Many automobile bodies spoke clearly of class distinction. There were two compartments—one in front for the chauffeur, who was responsible not only for driving the vehicle but for its correct mechanical functioning; and another at the rear for the passengers, who were divided from the driver by a glass partition to ensure privacy, but also provided with a speaking tube through which orders could be given (fig. 203).

Quite different was the fast owner-driven sports car, which presupposed that being in charge of a motor vehicle was in fact a pleasure in itself, more important than the mere business of being transported from one place to another. Some of these sporting cars were as large and heavy as the luxury limousines—good examples were the Prince Henry Vauxhalls, introduced in 1910. These had three-litre engines, increased by 1914 to four litres. But more significant for the future were the light sporting cars then called voiturettes. Perhaps the most distinguished pre-war example was the Bugatti Type 13 (fig. 202). This was introduced in the same year as the Prince Henry Vauxhall. Bugattis, like Vauxhalls, were often used for competition purposes, and the results of competition experience were in due course passed on to the customer, who was attracted as much by the mystique of speed as by the actual merits of the vehicles themselves.

Though early Bugattis and Prince Henry Vauxhalls continue to arouse the enthusiasm of automobile fanatics today, their historical importance is not in fact nearly as great as that of the Model-T Ford (fig. 205), which was introduced to the public in 1908. This was the first car designed for a mass market. It was based on two ideas. First, there was the design itself— simple, sturdy and easy to repair. The Model-T did not offer what its creator thought of as unnecessary refinements. Rather than add these, he preferred to cut manufacturing costs to put his product within reach of more people. He succeeded in this aim thanks to the assembly-line system introduced in his factories. The price of the Model-T was reduced in each successive year and motoring ceased to be a privileged occupation.

Some European manufacturers followed suit. The Morris Oxford became available in Britain in 1913 (fig. 206). The difference in design between this and the Model-T reflected the difference in physical conditions between the two countries where the vehicles were made. The Model-T was larger, sturdier and higher off the ground—it was meant to cope with the vastness of the American continent as well as the then rudimentary nature of many American roads. The Morris Oxford was smaller, because English lanes were narrow and twisting. The Morris also showed clear signs of the influence of existing luxury light cars, such as the Bugatti.

As a solution to a particular problem, the Model-T was so successful that it continued to be made in virtually unaltered form over a long period—until 1927. In nineteen years Ford made over 16 million examples. It is not in fact unusual, in automobile history, for a revolutionary design to continue virtually unaltered in this way. Other examples are the Citroën Light 15 Traction Avant and 2 cv (fig. 208), the Fiat Topolino (fig. 225), the Volkswagen (fig. 207) and the Austin Mini. These, which otherwise scarcely resemble one another, do have one important thing in common. They are all highly individual and out of the automotive mainstream. Each represents a radical rethinking of what an automobile can do, and how, therefore, it should be designed to fulfil its functions. The Citroën Traction Avant is the only one not specifically designed as a people's car—all the others were attempts to provide motoring for the many. Their designers, working within very narrow limits of cost, in each case were forced to rethink the way in which an automobile functioned from the beginning, almost without reference to previous solutions. The emphasis

206 Morris Oxford, 1912.

was on practicality, and frills were altogether lacking. The eccentricity of some of the designs—that of the 2 cv, for instance—won them a cult following. The slab-sided French car, with its willing but sluggish performance and soft springs, was so much the antithesis of anything glamorous that for that very reason it acquired a kind of reverse-chic.

It was also significant that the Fiat Topolino, and still more the Volkswagen, were 'made to order' cars, outside the normal current of development because they were attempts made by the Fascist dictatorships to redeem some of their promises to the mass of supporters who kept them in power. Pressure from above explains why their designers were able to go contrary to established automotive conventions. Professor Ferdinand Porsche, who had first worked for Austro-Daimler and later, under Weimar, had designed the famous sports cars produced by Mercedes, was in 1934 ordered by the Nazis to design

207 The Volkswagen Beetle, publicity photograph issued in 1976.

208 The original Citroën 2 cv, 1939.

and produce a people's car. Even if this order had not in fact fitted in with his own ambitions, he would have been forced to obey it. The project got under way immediately, but the finished product only started to trickle on to the market in 1938. Production resumed in 1945, and it was then that the Beetle captured the mass market. Over 15 million examples were sold. Its appeal was that it provided practical transport, but did not court comparisons with other cars—a Beetle owner could regard his vehicle as neither better nor worse than anything else he saw on the road—merely different. To choose to drive one was to make a clear-cut statement about one's attitude to the automobile, which was that it was a convenience, one to be kept in its place and not accorded undue importance.

The Beetle was nevertheless not quite without predecessors. In 1934 (a vintage year for automobile design) the Czech Tatraplan Type 77 was put on the market. This, though much larger than the Beetle, had a somewhat similar though more elegant shape, also governed by the presence of an air-cooled engine at the rear (fig. 209). The Beetle took technical ideas implicit in the Czech car to a logical conclusion. What was valid for the few might be even better for the many.

Another car which may have been somewhere in Ferdinand Porsche's mind when he elaborated his plans for the Beetle was the Chrysler Airflow Sedan, also introduced in 1934 (fig. 210). Chrysler had been making experiments with streamlining since 1927. These were not favoured by the company's business management and marketing side. What tipped the balance in the project's favour was the publication of Norman Bel Geddes's book *Horizons*. As a result Geddes was called in to help with the design of the bodywork, and also with the publicity—he proclaimed the Airflow to be 'the first sincere and authentic streamlined car', and this is still the reason why it tends to occupy a prominent place in design histories. But despite drawing record orders when it first appeared, the Airflow was a commercial failure. The reasons are likely to have been complex—failure to reach full production of the car quickly enough, defects in quality control, and perhaps the public's disappointment, after all the propaganda, that the car was not streamlined enough, the reason being that it had to be built around existing Chrysler engines.

It is nevertheless interesting to reflect that when 'outside' designers, like Geddes, have intervened in the automobile industry the results have seldom been successful. The car Gropius designed for the German firm of Adler, produced from 1929 to 1931, is remembered now chiefly because Gropius designed it and not for any very radical qualities of its own. The single exception is perhaps Raymond Loewy's connection with Studebaker, and especially his design for the Studebaker Avanti of 1962, a skilful adaptation of advanced design thinking to established Detroit conventions. One reason for the failure of other attempts by outsiders is that they too were cosmetic jobs, but much less cleverly executed. The designer was called in, in these circumstances, only when the fundamental engineering had been done.

Loewy is interesting for another reason. It must be obvious to anyone who reads his book *Industrial Design* (1979)—a more autobiographical work than the title might lead one to expect—that he has had a long and passionate love-affair with the automobile, in this resembling many of the other inhabitants of Western industrial societies. Loewy differed from most of his fellows in having the resources to give his ideas concrete form. He was constantly designing specials for his own use. For a motoring holiday in France at the height of the 'tail fin' era, he purchased a Cadillac and

209 Tatra Type 77, 1934.

altered both front and rear ends to suit his own taste. Loewy's specials are important not only because they often served as prototypes for vehicles which were later put into production but because they fit into a pattern of dream automobiles of various kinds which are an integral part of the story of the industry.

The spectrum is wide. One might, for example, dispute the claim of the Rolls Royce Phantom III of 1936 illustrated here (fig. 215) to be a dream car. When it was new, anyone with the money could have bought it; and its styling, though abreast of its times, was certainly not ahead of them. Yet, like all Rolls Royces of its period, it is not only individual, but unique. The customer ordered a chassis from Rolls and handed it over to one of a number of specialist coachbuilders, who constructed the bodywork to order, rather like a good tailor making a made-to-measure suit. A Rolls was a flamboyant statement of privilege, and it provided a goal which humbler automobile owners aspired to. So too, in an even more specialized way, did the Bugatti sports coupés of the middle and late 30s. These were often provided with bodywork of

210 Chrysler Airflow, 1934.

211 Chrysler Futura experimental car, 1955. from *Dream Cars*, 1981.

outstanding elegance designed by Ettore Bugatti's son Jean (fig. 216). Like Rolls Royce bodies, these were in no sense an industrial product but were laboriously handmade by craftsmen in small specialist workshops. Nor were the cars themselves in any sense maids-of-all-work. Their importance lay, first, in their influence on mass manufacturers as a source of ideas; and second, and more importantly, in their profound impact on the public imagination.

Ettore Bugatti himself was not immune to grandiose fantasies. His most ambitious effort was the huge Bugatti Royale (fig. 213), designed to be the car of kings. Six were made, and only three were ever sold. The Royale could be up to 22 ft. long, depending on what coachwork was fitted, and it had a wheelbase of 14 ft. 2 in. At a time when a Rolls Royce Phantom chassis sold for £1,900 ($9,500) in the United Kingdom, a chassis for a Royale cost £5,250 ($26,250).

The Royale in turn may have provided at least part of the inspiration for the fantastic automobiles occasionally faked up in Hollywood during the 20s and 30s for use in films (fig. 212). These were part of the whole larger-than-life atmosphere that the studios created, but they were also an accurate reflection of the secret aspirations of many movie-goers. The ancestry of the example illustrated clearly includes the Dusenberg, generally held to be the most luxurious American automobile, and driven by stars such as Clark Gable in 'real life'.

After the Second World War the American automobile industry took to producing occasional one-offs which were used to test public reaction to possible model changes, and which also served to generate publicity for a particular marque. Lincoln's Futura of 1955 (fig. 211) had a claustrophobic double-canopy, and featured a rear-mounted mike to pick up and amplify sounds made by the cars behind. Though the Futura, true to its name, was presented as a 'car of the future', it seems chiefly interesting now as an example of exaggerated 50s styling. These exaggerations were much condemned at the time, and have not won much sympathy since. Yet it must be admitted that they did sell automobiles. The company chiefly responsible for propagating the style was the leader of

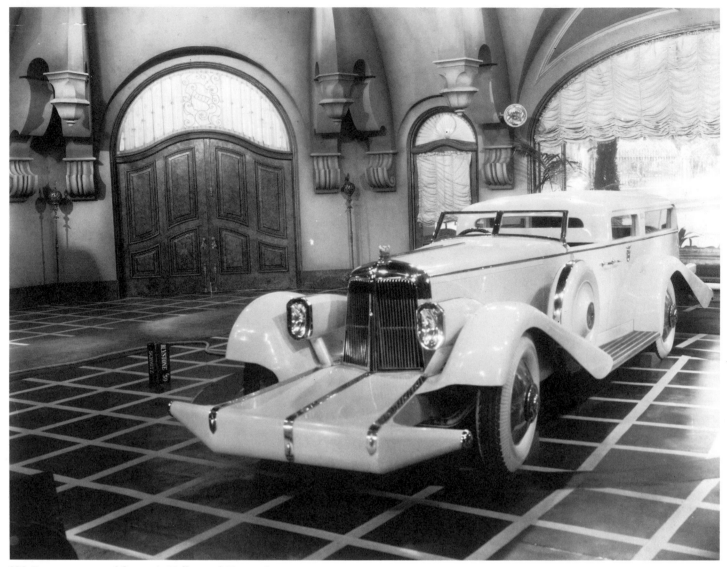

212 Dream car created for use in Hollywood 30s movies.

213 Bugatti Royale, Type 41, 1930.

214 1958 Buick Special Riviera.

215 (*above right*) Rolls Royce Phantom III, 1936.

216 (*right*) Bugatti Type 57 'S' Atlante coupé, mid-1930s.

the American auto-manufacturing industry, General Motors, and if any single designer was responsible it was undoubtedly Harley Earl.

Earl began designing cars in 1927, and his exuberant inventions of the 50s, like the Buick Special Riviera of 1958 (fig. 214), were the fruit of much thought, much experience of the industry, and much accumulated knowledge of the American consumer. He was the first to think of an automobile as a totally integral plastic shape, and it was under him that the custom was introduced of first modelling the bodywork full-scale in clay. An automobile became quite literally a piece of moving sculpture. He was as much interested in the symbolic aspects of design as he was in purely practical ones. The reason for the exaggerated tail-fins on many of his creations was that this detail suggested an aeroplane in flight—it is said to have been suggested in the first instance by the tail-booms of the P-38 aircraft used in the Second World War. The 1958 Buick also looked to more recent sources of inspiration—to the new jets of the period. Its wrap-round windscreen, so typical of the epoch, was an unnecessary technological flourish, but it, too, added to the impression of speed and communicated the idea that this was something absolutely up to the minute. The florid chrome trim was in harmony with the popular culture of the day, still hovering on the brink of investigation and definition, and not as yet self-conscious.

Automobiles of this sort did embody a very powerful myth, something soon to be distilled by Pop Art. It is no accident that one of the earliest works which can truly be called Pop is the British artist Richard Hamilton's *Hommage à Chrysler Corp*, painted in 1957. A series of studies for this (fig. 223) shows how carefully Hamilton explored the sexual qualities of Detroit design, and also the way in which these

qualities were presented to the public via advertisements. Detroit design was always heavily symbolic, and it was based on the notion that automobiles were purchased by men, and that the car was a female whom the man must master. European ideas of luxury were more restrained, tending to be based as much on class and snobbery as on sex.

One very important part of Detroit marketing strategy was the idea of an annual model change. The

217 Chevrolet Corvette, 1954.
218 Chevrolet Corvair, 1960.

man responsible for introducing this notion was also Harley Earl of General Motors, and it became part of the economics of the American automobile industry. Smaller or larger details of bodywork would be altered, to make this year's car visibly different from last year's. In addition to this, and even within a single range, the customer would be offered numerous options—in 1965 Chevrolet offered 46 models, 32 engines, 20 transmissions, 21 colours, 9 two-tone paint schemes and 400 accessories. The number of possible permutations was almost infinite. A series of pictures covering the period 1954 to 1968 shows some of the variations in Chevrolet design (figs. 217–22).

Yet it is also important to remember that in some ways the American automobiles of the 50s and 60s were remarkably similar to one another in basic type—they were large, with big, lazy engines, soft suspensions and extremely low-geared steering.

219 Chevrolet Corvette, 1961.
220 Chevrolet Impala sport sedan, 1964.

221 Chevrolet Corvette Stingray coupé, 1965.
222 Chevrolet Chevelle Malibu, 1968.

European drivers complained of their unwieldiness and lack of road-holding, while admiring their often fierce acceleration away from traffic lights. They were, however, very precisely fitted to what was then the American way of life. They were comfortable for long journeys on wide roads, yet able to cope with occasional rough tracks, and they were reliable because not too highly stressed mechanically. Americans looked on their cars as total environments, soothing and womb-like.

The post-war Detroit designers have often been condemned for the wastefulness and lack of logic to be found in their creations, as well as for tastelessness. Taste is always a matter of opinion—Detroit design, for better or worse, was an accurate reflection of a whole culture, of its dreams as well as of its practical needs. In fact, it can even be argued that Detroit's logic, in a certain sense, was too accurate for the industry's long-term good. American designers found great difficulty in believing that the situation to which they had adapted themselves and their product could ever be altered in any fundamental way. The crisis which overtook the American automobile industry in the 70s, and which remains with it today, was due not so much to an arbitrary change of preference among its customers as to the sudden shortage of energy. This led to strict speed limits, and to an emphasis, hitherto unknown in America, on the need to conserve fuel.

These restrictions, in turn, coincided with a shift of attitudes. The automobiles had become ordinary—Americans, like their European counterparts, were ceasing to be car-worshippers. Thanks to the success of cars like the Beetle and the 2 cv, European manufacturers were perhaps quicker to spot this fact than their American rivals. The development of a small European family saloon was by the early 80s not just a matter of massive investment but a strict application of design logic in a way which tended to rule out the old romantic considerations.

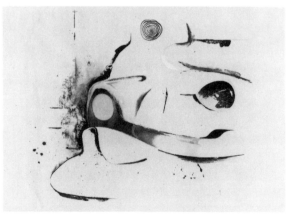

223 Richard Hamilton, study for *Hommage à Chrysler Corp*. Private collection.
224 (*opposite*) Early design sketch for the Fiat Panda.

An Automobile Case History

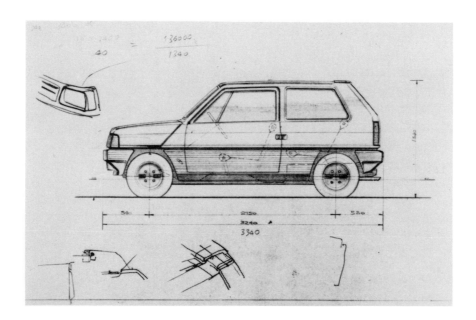

An automobile is clearly one of the most complex objects that an industrial designer can be called upon to produce. It can also be argued that successful design for the broadest possible market is also one of the most stringent tests of a designer's ability. Bearing these two points in mind, a particularly informative case history is that of the Fiat Panda, the small compact vehicle launched by the great Italian manufacturer in 1980. The factors which governed the design are perhaps the more visible because Fiat called in (and subsequently gave fullest credit to) a major Italian design studio—ItalDesign, headed by Giorgetto Giugiaro. The design is therefore largely attributable

to a single individual, which is by no means always so in this area of manufacture.

Where the Panda is concerned, one can speak of two traditions coming together, and the decisive rejection of a third—the tradition of Detroit and Harley Earl. The Panda is an extreme statement of the current European thinking about automobiles. Its immediate ancestors are, with one important exception, cars previously produced by Fiat itself. It is in direct line of descent from the Fiat 500 Topolino (fig. 225), already referred to in the last chapter; the Fiat 600 of 1955; the Fiat 850 of 1964, and—still current at the time when the Panda was introduced—the Fiat 126 (fig. 226) and

127. The other automobile which exercised a decisive influence on design thinking for the Panda was the Citroën 2cv (fig. 208), not by reason of its shape, method of manufacture or characteristic mechanical features, but because it was a roomy, simply constructed vehicle powered by a comparatively tiny engine—and, in addition, it was notably light for its size. The other design tradition was one which had been evolving over a number of years in Giugiaro's own studios.

Both Fiat and Giugiaro agree that the brief the designer was given for the new Panda was in many respects extremely free—there were no restrictions as to what the finished result should look like. But it was tightly restrictive in others. The limitations were those of overall weight, manufacturing cost, and the type of engine to be used. Fiat wanted a car which was as light as their little 126 (often regarded as primarily a town car), and which cost no more to manufacture. The air-cooled 126 engine was to be used—but Giugiaro also surmised from the beginning that Fiat

would also want to employ their more powerful 127 engine. The fact that this was water-cooled provided an additional complication.

The rest of the design sprang from a single fundamental idea, one which, though by no means new, is still surprisingly revolutionary in terms of the world automobile industry. The Panda was to be designed 'inside out', starting from the driver and passengers. Fiat specified a full five-seater vehicle, the dimensions to be based on the standard 'manikin' devised by the Italian SAE (Society of Motor Engineers). The manikin is a full-sized jointed dummy used as an objective measure of human proportion. Felice Cornacchia, the head of Project Development at Fiat, points out that this dummy is based on notions which can be traced back to Leonardo da Vinci.

From the beginning the Panda seems to have provided the basis for an unusually smooth collaboration between Fiat and their chosen designer. Cornacchia describes it as an 'intelligent' as opposed to a

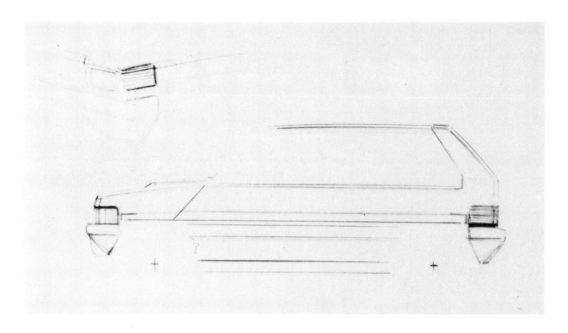

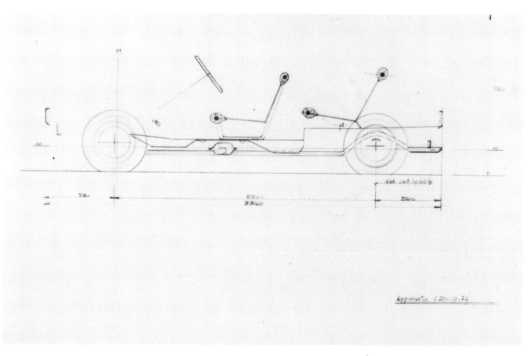

225 (*opposite below*) Fiat 500 Topolino, 1938.

226 (*opposite above*) Fiat 126, 1972.

227 (*left*) Giugaro's first rough sketch for the Panda.

228 (*below*) First concept for the interior of the Panda, July 1976.

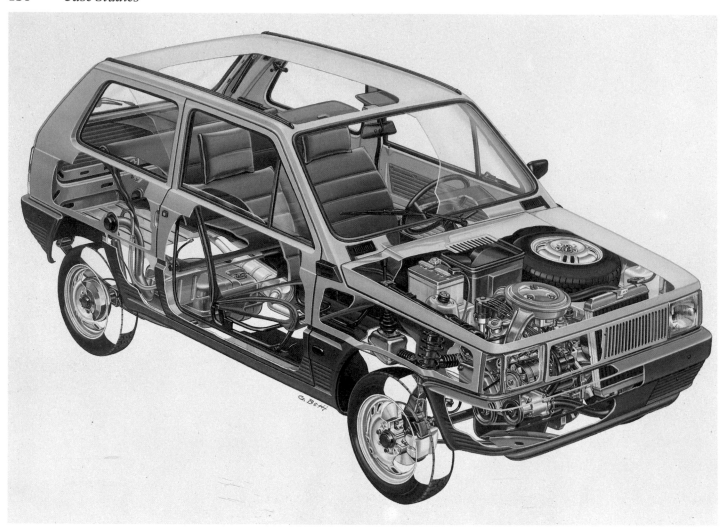

'luxurious' automobile, and also comments that it is 'the minimum we can describe as being a car in the full sense'. Giugiaro remarks: 'I think the Panda is the most amusing project I have ever been given.' In his view, it finally gave the lie to the notion of the automobile as a status symbol.

Work on the project, first code-named 'Rustica' or 'rustic'—a significant choice of adjective—started late in the summer of 1976. Giugiaro's own original ambition was to be, not a designer, but a fine artist, and the shape of the vehicle first crystallized in some rough sketches he made while on vacation (fig. 227). This was followed on his return by studies of details, including some for the passenger compartment, with deliberately minimal solutions for items such as the seating (fig. 228), in keeping with the rest of the concept. The seats are basically thin padding slung over a tubular framework (fig. 235)—which does not rule out a good deal of ingenuity. The back seat is adjustable to seven different positions, and can even be used to make a bed.

These studies were followed by a full-scale plaster mock-up, almost indistinguishable in a photograph from the real thing. This mock-up was made in two versions—the one preferred by the designer, and a more conservative version closer to the current Fiat production (fig. 230). There were also full-scale interior mock-ups to show details (figs 231, 232). By September 1977, ItalDesign had produced a rolling prototype, with a chassis capable of accommodating various engines. About 20 of these were delivered to Fiat, and were used for checks of various kinds, in the laboratory and on the road. One concern, since the car was so light, was the overall strength of the structure.

229 (*left*) Fiat Panda, exploded view.
230 Two mock-ups for the Panda—the one on the left was preferred. In the middle is the Fiat 126.

231, 232 Interior mock-ups for the Fiat Panda.

When the Zero, as it was now called, passed these tests with flying colours, ItalDesign were also asked to design the actual tools necessary for industrial production.

At this point it is useful to return to a direct comparison with the Citroën 2 cv. Felice Cornacchia speaks of the 2 cv as an 'artisanal' rather than an 'industrial' product. What he means by this is that the simplicity of the Citroën's basic shapes has to paid for by the number of welds which are necessary to assemble it. Similarly the extreme lightness of structure is paid for, in his view, by the exaggeratedly supple suspension which avoids putting too much stress on the framework. The Panda, he claims, is both simpler and also cheaper to produce, in terms of a sophisticated production line.

The car certainly compares favourably in production terms with the other small models in the Fiat range. The number of body components has been reduced by about 18 per cent, and the number of spot welds cut by 28 per cent.

It is a sign of the times—and of a change in industrial philosophy—that great attention was paid not only to ease of manufacture but to durability. Fiat claim that the Panda is designed to have a useful life of

233 (*above left*) Volkswagen Golf GTi, 1983.
234 (*above*) The Panda 4-door—an unrealized concept.
235 (*left*) First proposal for the Panda front seat.

10 years and 100,000 kilometres. In the near future it will improve to 10 years and 150,000 kilometres. The life-span remains the same though the possible distance travelled increases because of the still intractable problem of corrosion, which affects all motor vehicles.

The notion of increased durability has affected the Panda's styling, not merely overall, in the search for something sufficiently individual not to date too badly, but also in specific details. One of these is the treatment of the body sides below waist level. These are coated with a grey polyester substance which has 10 times as much resistance to damage as ordinary paint. This strip protects against stone-chipping and other minor damage. It also has the advantage of reducing the apparent height of the Panda—the high, boxy shape is increasingly typical of Giugiaro's designs and is found in less exaggerated form in his treatment of the Volkswagen Golf (fig. 233), which represented his real breakthrough into the mass market. Height and boxiness reflect his conviction that in everyday cars—the workhorses of our society—carrying capacity and roominess can be too lightly sacrificed to aerodynamics. The Panda can thus be seen as an extreme reaction against the Streamline Style, and also against its successor, favoured by Giugiaro for various glamorous sports cars, the so-called 'flying wedge'.

The Panda's story does not end with a successful launch in 1980. Once launched it has, like most successful automobiles, taken on a life of its own—not least because its represents such a huge investment for the parent company. Some modifications come from ItalDesign, some from Fiat's own Styling Centre (a department directly responsible to Cornacchia). One or two modifications were imposed because the original was no longer considered satisfactory after more prolonged experience of the car's behaviour in use. In the case of the Panda both the suspension system (particularly the rear suspension) and the brakes have been improved since the car was introduced. The problem with the rear suspension is an example of the compromises inherent in nearly all automobile designs—Giugiaro had been anxious to get a wide, totally flat loading area at the rear, and this, together with considerations of cost, ruled out an independent rear suspension.

More luxurious versions of the Panda, with less spartan but perhaps less ingenious fittings, have been introduced since the launch, and there is still a constant search for attractive variants which may appeal to one or another specialized segment of the market. Fiat reckon they have to sell 15,000 examples a year of any of these variants, in order to justify the expense of putting it on the market.

236 The Panda Nero—an unrealized design for a sports version.

237 The 4 × 4 Panda Strip, 1982.

Even when he was first designing the car, Giugiaro envisaged, though perhaps in this case more for his own amusement than for any practical purpose, both a four-door version (fig. 234), and an idea for a Panda sports car which might have looked a little like a pick-up truck (fig. 236). This latter notion, however, did find a practical future in the 4 × 4 Panda Strip which is just about to be introduced at the moment of writing (fig. 237). The 4 × 4 Strip is a sporty, slightly Jeep-like vehicle with a new type of four-wheel drive, created and patented by ItalDesign and sold by them to Fiat— an example of engineering as well as pure design capability within a big independent design studio. This new variant is intended as a 'fun car' for the younger generation—a somewhat more practical successor to the beach buggy and the Mini-Moke.

Despite this constant variation from the original concept (something Henry Ford would not have approved of), the Panda does in fact have quite a lot in common not only with its immediate Fiat, Citroën and Volkswagen exemplars, but with the Model-T. It illustrates the positive aspects of contemporary industrial design because it is in its own way and in its own time just as much of a radical solution as Henry Ford's brainchild. Its emphasis on practicality rather than stylistic considerations reflects a new spirit in the contemporary automobile industry, as well as stressing the difference between the European and American design tradition in this field. It is contemporary in more ways than one—a preference for the minimal can be seen as a matter of aesthetic decision, as sound commercial common sense, and as an instinctive response to increasingly harsh economic facts.

238 Kitchen's Annular Biplane, *c.* 1910.

Other Kinds of Transportation

The paradox about the automobile, from a designer's point of view, is that the brief to create such a thing is necessarily unspecific. Even where conditions are laid down, there is a good deal of leeway in how they can be interpreted. It is much easier to produce a successful result when the requirements are wholly unambiguous. Franco Campo and Carlo Graffi, two architects working in Turin, were asked in the early 50s to produce a mobile touring exhibition advertising compressed bottled gas. The result (fig. 239) is a design which still looks extremely elegant two decades later. The bodywork they devised not only looks good in 'road' form, it is extremely practical. All that needs to

be done in order to convert the vehicle into an exhibition stand is to swing up the top panels along its sides so that they form canopies, and swing down the lower ones so that they make gangways. The success of this design springs largely from the application of a rigorous discipline to a narrowly defined problem.

Vehicles designed to transport passengers not in threes and fours but in groups also have to be designed to a tighter brief than private vehicles. The London omnibus, still a brilliantly successful solution to this problem, was not the brain-child of an individual or of a design team, but is something which evolved gradually. The two-decker bus is a descendant of the

239 (*top left*) Italian touring exhibition advertising compressed bottled gas, early 1950s. Coachwork by Macchi, Varese on a Lancia chassis. Designers Franco Campo and Carlo Graffi.

240 (*top right*) N.S. type London bus, 1925.

241 (*left*) A London bus in 1973 with advertising livery.

242 (*right*) Greyhound bus designed by Raymond Loewy, late 1940s.

circus-wagon, and still retains links with the past in the shape of the lively advertisements which are often seen on its sides. An example of 1925 (fig. 240) shows it at a midway stage in its evolution. The top deck, which was at first open, is by this time covered over, but the coachwork still makes the point that the vehicle is built up of quite separate parts—two decks, one on top of the other, a compartment for the engine in front, and a curving staircase tacked on behind. By 1973 (fig. 241) the design has become, as far as possible, a unified slab, with four wheels underneath.

Given the sort of task it was expected to perform, it was neither possible nor indeed plausible to streamline the London bus. Designing for the long-distance Greyhound Bus Company in America, Raymond Loewy did introduce some elements of the Streamlined Style (fig. 242), though the basic shape remains

recalcitrantly slab-sided so as to accommodate as many fare-paying passengers as possible. The most notable features of Loewy's design are symbolic ones. He redesigned the company's emblem, the greyhound. In 1933, when he received his first Greyhound assignment, Loewy told Orville S. Caesar, Greyhound's chief executive, that he thought the silhouette then shown on their buses looked like 'a fat mongrel'. The image Loewy developed played a key role in providing the company with an identity immediately recognizable to the travelling public.

Railway engines, which came into existence long before automobiles, were for a long time entirely in the hands of the engineers. Indeed, it was not until Loewy's generation that the design profession was called in to play a part in reshaping the way they looked. The British Great Northern Railway built the

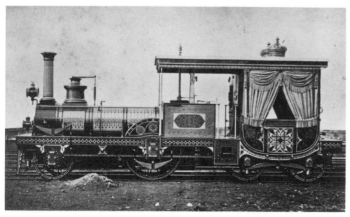

243 Stirling 8-foot single engine no. 1, built for the Great Northern railway, 1870.

244 Coronation 4–4–0 engine built for the London and North-Western railway, 1910.

245 Locomotive built *c.* 1860 for the Khedive of Egypt, designed by Sir Matthew Digby Wyatt.

Stirling 8 ft. Single Engine No. 1 which is illustrated (fig. 243) in 1870. Its basic forms, however, can be traced back to a much earlier time. Apart from the extra pair of wheels in front, this engine is still very close to the *Jenny Lind*, built for the London, Brighton and South Coast Railway in 1847. The *ad hoc* design of the very earliest locomotives, such as George Stephenson's *Rocket* of 1829, had by that time been cleaned up and rationalized. There was already a search for a simple, uncluttered look, and this had been pushed still further by the time the London and North Western 'Coronation' type engines were built in 1910 (fig. 244). One change from 1870 is that the whole driving motion is now inside the frame, which makes the design much tidier. To this extent at least, those responsible for designing railway engines were conscious of the way they looked as well as of mere efficiency of function.

Design in the fully conscious sense was only applied to early railway engines on very special occasions, and then the results tended to be rather odd. Around 1860, Robert Stephenson of Darlington built a state locomotive for the Khedive of Egypt (fig. 245), and this was designed by a fashionable architect, Sir Matthew Digby Wyatt, builder of what has been called 'the most elegant Second Empire mansion in London', Alford House in Prince's Gate. Wyatt knew something about railways—he collaborated with Brunel, and built Paddington Station for him—but here we see him imposing completely extraneous details on the engineers' conception—for example, turning the chimney into a classical column.

American locomotives were quite markedly different in design from British ones because of a great difference in conditions. A typical early example of 1857 (fig. 1) is fitted with a cow-catcher to deflect objects on the line and prevent them getting under the

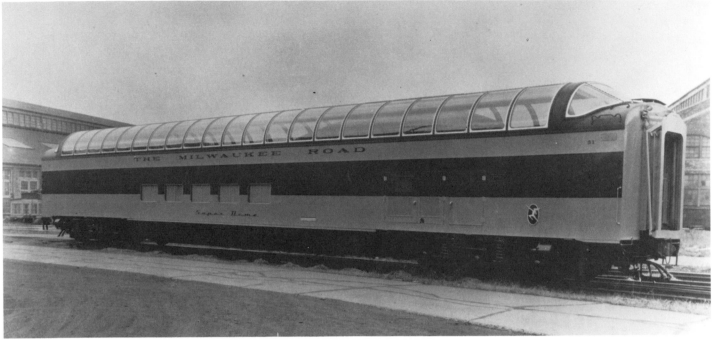

246 (*above*) Two-deck 'Super-Dome' observation car built for the Milwaukee railroad by Pullman-Standard, mid-1950s.
247 (*below*) British Rail's Advanced Passenger Train (APT), 1982.

wheels, and a warning-bell—both necessary because the American railways ran, for much of the time, through open, unfenced country. The cab is enclosed, since the driver had to go long distances in often severe weather conditions; the chimney is tall and fitted with a spark-catcher because wood was usually burnt, not coal; and the engineering layout is deliberately open, so that if a breakdown occurred it was easy for the driver—significantly called the 'engineer' in the United States—to make repairs along the way.

When designers who had no railway background began to be employed by the great American railway companies, they often met fierce resistance from the engineering staff. This was the case when Raymond Loewy was taken on by the Pennsylvania Railroad. The Penn's engineers loved the iron horse the way it was, with many of its working parts exposed. In fact, their fascination with its workings was for many of them a strong reason for joining the railroad in the first place. Loewy's determination to conceal what was happening as the locomotive moved was for them a kind of heresy, and they were unmoved by the argument that streamlining saved energy. Nor is it quite certain that Loewy's motives were purely scientific in streamlining locomotives in this way, though he subjected his models to wind-tunnel tests before the concept was finalized, as a sort of objective justification of what he proposed, and then (following Detroit's already established practice with auto- mobiles) had a full-scale mock-up sculptured in clay, to give an idea of what the finished result would be like.

But streamlining was also an aesthetic choice, and the symbolic and metaphoric aspects of the task were clearly very much in Loewy's mind when he designed the S-1 locomotive (fig. 3). As he himself points out,

248 'Sky top' rear observation car built for the Milwaukee Railroad's 'Hiawatha' train, 1948.

249 DE-III Trainset built for the Netherlands State Railways, 1935.

the silhouette of an aircraft which appears in his first sketch for the project is an amazing anticipation of the Concorde. It was because they were conceived with imaginative force, not merely through a process of objective logic, that the S-1 and the other streamlined locomotives which were its contemporaries had a huge impact on the public. They gave travelling by rail renewed glamour, a quality the railroad had begun to lose in the face of competition from automobiles and aircraft.

Just as important in putting the message of glamour across were the luxurious observation cars and other facilities which were added to trains to attract passengers. The width of the American gauge, and the absence of low bridges, made vast double-decker cars possible on some routes (fig. 246). The search for visible glamour did, however, lead to the creation of designs which were only doubtfully functional, like the 'Sky Top' observation car (fig. 248) put into service on the Hiawatha train of the Milwaukee Railroad in 1948. Here the opulent massiveness of the design

resembles that of American automobiles at the same period. It is incidentally amusing to note that the seats themselves faced inwards!

The logical conclusion to streamlining railway engines was to streamline the whole train, making it a unit, though this gave less flexibility of operation as the carriages used were not interchangeable with others of standard design. One of the best and most practical early designs of this type was introduced on the Netherlands State Railways in 1935 (fig. 249). Astonishingly, it was designed, and 40 train-sets were constructed, within the brief span of 15 months. The trains were powered by German Maybach diesel-electric engines, and the carriage bodies were made of light sections and tubes, with no separate underframe. Doors, and details such as luggage-racks, were made of aluminium alloy. These train-sets do not look dated even today, and indeed, though not designed to travel at tremendously high speeds—on the schedules they were meant to maintain maximum speed was no more than 62 m.p.h.—they bear quite a marked resemblance to British Rail's much-touted Advanced Passenger Train (fig. 247), which is still being developed and tested as this is being written. The special feature of the latter is the tilting mechanism to ensure passenger comfort as the train traverses curves at high speed—up to 125 m.p.h.

Where sea travel was concerned, nineteenth-century engineers sought not only for technical improvement as such, but for an increase in size which would allow much greater comfort to passengers, as well as an increase in payload. New ideas were tested on the profitable transatlantic route. The first steamship built to make regular Atlantic crossings was Isambard Kingdom Brunel's *Great Western* (fig. 250), which made her maiden voyage in 1838. She was wooden built and a paddle-steamer—two features

250 (*opposite above*) S.S. *Great Western*, 1837.
251 (*opposite below*) S.S. *Great Eastern*, 1858.
252 (*above*) T.S. *Titanic*, 1911.

which were soon to be superseded—but the most important feature of her design was the fact that she was sufficiently economical in her use of fuel to arrive in New York with coal to spare. Much bigger, and technically much more advanced, was the *Great Eastern*, also designed by Brunel (fig. 251). She was laid down in 1854 but launched only in 1858. By the standards of the time she was immense—18,914 tons when the biggest ships then afloat were then under 5,000. She was powered by a propellor in addition to her paddle-wheels, and had a cellular double-bottom. She could carry 4,000 passengers plus 6,000 tons of cargo. Brunel meant her to be used on the long routes to India and Australia, where here capacity would have been an advantage, but like the *Great Western* she was put in service on the North Atlantic route and failed to generate enough business to pay the high costs of running her. She finished her career not as a passenger-carrying vessel but as a cable-layer.

253 The Wright Brothers make their first flight at Kittyhawk, 1903.

254 (*opposite left*) A Blériot monoplane in R.A.F. markings, pre-1914.

255 (*opposite right*) The Sopwith Camel, 1918.

Another failure, of a far more tragic and spectacular sort, was the White Star liner *Titanic* (fig. 252), sunk by an iceberg on her maiden voyage to New York with the loss of nearly 1,500 lives. By her time the design of large passenger ships had apparently taken several steps forward. She was of 46,323 tons, and her hull was divided into 16 watertight compartments—contemporaries regarded her as virtually unsinkable. On the other hand, she did not carry enough life-boats, and her large and magnificent public spaces—a ship of her type was now fully the equivalent of a luxury hotel ashore—made her vulnerable. In fact she typified a design conflict which the builders of luxury liners have never quite resolved—between passenger comfort and beauty of appearance on the one hand, and absolute standards of safety on the other. It has recently become the custom to employ intermediary 'design engineers' to resolve conflicts of this sort in both large passenger ships and passenger aircraft.

The mode of transportation where the whole question of mechanical safety and efficiency is paramount is, however, not travel by sea but travel by air. The story of flight belongs wholly to our own century, starting with Orville Wright's epoch-making effort in the Wright brothers' Flyer at Kittyhawk, on 17 December 1903 (fig. 253). The problem the Wrights were the first to solve was not merely that of getting a heavier-than-air machine to fly, but the problem of controlling it when it was up. Many early aircraft proposed what now seem to us the strangest solutions to these two problems (fig. 238). In order to take man into the air, the Wrights had to abandon one notion which had obsessed earlier experimenters—this was that a heavier-than-air flying machine ought to be a direct imitation of a bird. The flyer looks more like a box-kite. But it did in addition employ the important principle of wing-camber, where the upper surface of the wing is more deeply curved than the lower and thus creates lift as the airstream divides to flow both over and under the wing. But while this solved the business of getting the craft into the air, it still had to be moved under control through three dimensions. The inventive individualists who constructed the first aircraft solved these problems by trial and error, as there was no accumulated body of information to draw upon. Once the basic principles had been grasped, progress was fairly rapid. The type of monoplane designed by Louis Blériot (fig. 254), and used by him to cross the Channel in 1909, already marked a great step forward.

Yet technically the aeroplane did not truly come of age until it was forced to do so by the First World War. A fighter aircraft like the British Sopwith Camel (fig. 255), which brought about the virtual eclipse of the German air force in 1917, was extraordinarily agile in the hands of a moderately skilled pilot. It had to be, as superior manoeuvrability gave him a greatly increased chance of staying alive.

The next stage, after the war, was to develop the aeroplane so that it could be used for carrying passengers on a regular basis. The first services were flown by ex-military aircraft modified for civilian use. When specially designed aircraft began to be introduced, the monoplane gradually drove out the biplane, and all-metal construction replaced fabric and wood. The Fokker Trimotor (fig. 256) was perhaps the most successful airliner of the late 20s. This was at the half-way stage—a high-wing monoplane still using a lot of fabric.

This in turn was displaced by the American-designed Douglas DC-3, shown here in an evocative photograph by Margaret Bourke-White which captures the feeling of excitement still generated by

aviation in the 30s (fig. 258). The DC-3 first flew in December 1935, and by the time production ceased in 1946 about 13,000 examples had been built for both civil and military purposes. It had an all-metal structure, a low as opposed to a high wing, and a retractable undercarriage. It also had variable-pitch propellors, which gave much greater economy. As far as passengers were concerned the DC-3 offered almost everything that airline passengers expect today, with the important exception of pressurization.

In the 30s the long-haul intercontinental flights were not the business of the DC-3, which had too short a range, but of the great flying boats, a now extinct breed. These vast roomy aircraft seemed to offer great advantages—it was easy to find suitable places for them to take off and land; they were luxuriously roomy, and the metaphor of a 'flying ship' appealed enormously to both passengers and to the men responsible for running the great airlines. Pan American Airways, founded in 1927, built its fortunes on the glamorous China Clippers it flew on the transpacific route. British Imperial Airways enjoyed a similar success with their Empire boats, predecessors

256 Fokker Trimotor, 1932–5.

257 Short Sunderland Flying Boat, 1950.

258 T.W.A. DC-3 mailplanes photographed by Margaret Bourke-White in the 1930s.

of the Short Sunderlands used for patrol duties during the Second World War (fig. 257). The future of the flying-boat still seemed so bright that a good deal of effort was put into developing new and even bigger types during and just after the war. One was Howard Hughes's immense 'Spruce Goose' (the H-2 Hercules). It made its maiden and only flight in 1947—one mile at a height of 50 feet with Hughes himself at the controls. Another was the British Saunders Roe Princess (fig. 259). The prototype first flew as late as 1952, when it had already been announced that flying-boat services were to be abandoned.

The fate of the Spruce Goose and the Princess indicates that technological and design development does not always have a completely logical or expected pattern even when, as with designing aircraft, the terms seem to be narrowly defined by the physical problems to be solved. What the designers of these two aircraft did not foresee was that the flying boat itself would turn out to be a dead end in the history of aviation development.

The Spruce Goose is interesting not only for its form but because it was the largest aeroplane ever, yet built of plywood—a seeming reversion to earlier technology prompted by wartime shortage of metal. A far more successful and celebrated aircraft constructed of this material was the British fighter-bomber, the Mosquito (fig. 260). The fuselage of the aircraft was a moulded plywood shell constructed in two halves. In the short term the Mosquito must be accounted a

considerable design success—a brilliant piece of problem-solving. But when the special conditions which prompted its manufacture evaporated, the techniques used did not continue to develop.

More significant in terms of aviation history, but nevertheless something with a question mark against it, was the first jet-engined airliner—the De Havilland Comet I (fig. 261). The first jet aircraft was an experimental German Heinkel, first flown in 1939. During the war the jet engine—a different type from that used by the Germans—was successfully developed by the British under Sir Frank Whittle. It was only logical after the war to plan a civilian passenger-carrying aircraft which made use of the technological

259 Saunders Roe Princess Flying Boat, 1952.

261 Comet jet-airliner by De Havilland, first flown July 1949.

260 Mosquito Fighter-Bomber, 1942.

lead the British had built up, and the first prototype of the Comet flew in July 1949. The aircraft was actually put into commercial service in 1952, and attracted a huge amount of enthusiasm both from experts within the industry and from the general public. Then, abruptly, its reputation was shattered by two tragic and at first mysterious disasters above the Mediterranean in January and April 1954. Rigorous investigation, which entailed lifting the remains of the aircraft from the sea-bed, revealed that the Comet was subject to certain types of stress, which induced metal fatigue. The British lost the leadership the Comet had brought, and their aircraft construction industry suffered a setback from which it has not yet recovered.

A somewhat different kind of ambiguity attends the reputation of what is now the most advanced passenger-carrying aircraft in the world—the supersonic Anglo-French Concorde (figs. 262, 263). Supersonic flight continues to arouse widespread opposition from conservationists and amenity groups because of nuisance from noise and actual physical damage to property. Nevertheless, Concorde must be counted a great success from the technological point of view—it fulfils the aim of its designers, carrying passengers at supersonic speeds regularly and safely. What remains in doubt is its economic viability. Only 16 Concordes have been built, and it is clear that they will never repay their huge development costs. It is even doubtful if Concorde is economically viable on a day-to-day basis.

The locomotives, ships and aircraft surveyed briefly in this chapter present both the engineer and the industrial designer with some of their most complex tasks, not made any less complex by the fact that these individuals, when their functions can be separated, are often in conflict with one another—the engineer tending to approach the task from the standpoint of mechanical efficiency, the designer from that of passenger comfort and sometimes also of appearance. It is perhaps not surprising to note how many of the pioneering examples I have chosen rate as failures.

Failure can be of several different kinds. It can happen that the engineer's reach outstrips his actual

262, 263 Concorde supersonic jet-airliner, in regular service in 1976.

technological capability. This clearly happened in the case of the Comet, and (with reservations because sister ships continued to operate successfully) one can interpret the *Titanic* disaster in the same way.

A commoner kind of failure is rooted in the conceptual stage. The designer—using this term in a general sense to designate the man with overall responsibility for creating a particular object— usually ignores the past at his peril, but can also be blinkered by his knowledge of it. He then produces something which is undoubtedly in advance of what precedes it, but which is nevertheless irrelevant to the prevailing conditions. The last of the great flying boats, like the Saunders Roe Princess, were doomed because large airports on land were being built, and these made it possible to operate much bigger land-based aircraft. A land-based aircraft with retractable wheels was more economical to operate, and also

faster, because it did not have to make the compromises forced by the need to land, and float, on water.

Finally, there is the sort of failure which occurs when vaulting technological imagination cuts loose from practical commercial considerations. Brunel's *Great Eastern* fell into this category, and it is likely that Concorde does so too.

Good design—which is sometimes in this case no more than good packaging—can at least momentarily rescue an established technology which is in danger of becoming obsolescent. Raymond Loewy performed this service for the American railway system in the late 30s, but was unable to halt its eventual decline in the face of increasing competition from automobiles and air travel. It we confront Loewy's spectacular S-1 locomotive—still an image of mechanical grace and power—with its contemporary the DC-3, there is no doubt as to which is the more significant. The DC-3 represented a coming together of a number of very different factors, technological and economic. One could say, for example, that the design was predictated on the availability of the variable-pitch propellor and a retractable undercarriage. If these were to be properly used aircraft had to be designed for higher wing-loadings and a lower drag-coefficient. This led to the sleek, low-wing shape which made the DC-3 such a contrast to an earlier generation of airliners such as the Fokker Trimotor and its all-metal successor and rival the Ford Trimotor. Changing the shape of airliners also made them much more economical, and it has been said that the DC-3 was 'the first aircraft capable of supporting itself economically as well as aerodynamically'. That is, it successfully solved all the problems which its designers set themselves, and did so so thoroughly that the solutions remained valid for at least a decade.

Transport design has now become so complex that it is difficult to attribute individual responsibility. Is the true designer the engineer? Or is he someone who is called in afterwards to titivate—or merely edit—what the engineer has produced? Our concept of design often seems to exclude major inventions, like the jet engine, yet it is the availability of these as power units which governs virtually every other feature of a modern aircraft. In major design projects the inventor, the engineer and what we designate as 'the designer' in a narrow sense have functions which overlap and interlock in an inextricable way.

Places Which Move

It has been said that all conveyances, on land, sea or in the air, can conveniently be divided into two basic types—simple transporters, or places which move. To a quite surprising extent the whole history of modern design is concerned in one way or another with the latter. At first sight, the problems they set seem easy enough to define. Places which move are forced by their very nature to perform more than one function. And space is at a premium.

One might take a caravan or a motorized camper as an example. The first purpose-built holiday caravan (fig. 264) was constructed in 1885, for a writer of popular boys' stories, Dr W. Gordon-Stables, R.N. He

designed it himself, and the builders were the Bristol Waggon and Carriage Works Ltd., whose normal business was making Pullman cars for the railways. Dr Gordon-Stables wanted to incorporate a full range of middle-class comforts in what had previously been a conveyance reserved for poor tinkers and gypsies. He was moderately successful in doing so. Among the items he managed to include were a sofa which turned into a bed, a bookcase, a mahogany chiffonier with a large mirror over it, a rack for hats and gloves and a marble washstand. The doctor left his wife at home, and was accompanied on his travels by a valet and a coachman. The valet slept in the caravan's kitchen on two long doormats with a cork mattress on top. The coachman, plus the horses, put up at a local inn.

These echoes of class distinction cannot disguise the fact that Gordon-Stables's design is the direct ancestor of the caravans and motorized campers we know today—for example the compact and practical Devon Moonraker which is built on a Volkswagen Transporter chassis (fig. 265). Here, too, a very adequate standard of comfort is provided within a very small space. It may seem that the essential task in designing all places which move is therefore to put a quart into a pint pot. But in fact space restrictions are seldom as stringent as these two examples suggest. A railway carriage provides a great deal more elbowroom. And unless it is meant to be used as a sleeping car it also confronts the designer with a rather different set of problems.

A railway carriage is not a home-from-home but a public place which happens to be on the move. While the problem of space is not acute, the problem of proportion is, since carriages are always long and narrow. The designer has the choice of frankly acknowledging this, or of trying to disguise it by dividing the area up. For a long time different solutions were adopted in the United States and in Britain. American rolling-stock was open plan, while the British had a stubborn preference for compartments. An American Pullman parlour car of 1876 (fig. 266) implies a certain gregariousness, though within a well-defined income bracket. The two rows of armchairs are fixed in place, but also swivel. This allows the passenger occupying one of them to shift position—either to ease his limbs, to turn to look at the view, or to engage in conversation with his neighbour. The image in the designer's mind seems to have been that of a gentleman's club, as the armchairs in particular bear witness.

A de luxe coach of the late 1940s, designed for the New York, New Haven and Hartford Railway, shows a distinct change of social as well as of visual style. Visually it could be described as post-Bauhaus rather than neo-baroque. The passengers sit in a firmly fixed relationship to one another, and they are crowded together more (fig. 267). A journey by rail is no longer a leisurely and pleasurable thing—it is a mundane but hopefully comfortable means of commuting between home and office. Another conspicuous difference between the two designs is a much greater concern for cleanliness—the seats are framed in metal, and the floors are composition not carpet.

Conviviality on the railroad did not go out with the nineteenth century. In particular, designers put some of their best thoughts into bar-cars and restaurant cars, coming up with solutions of bewildering variety. Some put the stress on simplicity and ease of

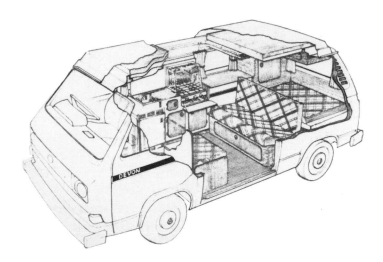

264 (*opposite*) 'The Wanderer'—the first purpose-built caravan for holiday use, 1885.
265 The Devon 'Moonraker' on a Volkswagen Transporter chassis, built by Devon Conversions Ltd., Exeter.

maintenance. An American restaurant car of the late 40s, designed for the Pere Marquette Railway, is simple and well thought out, with a diagonal seating arrangement to make the best use of the narrow space (fig. 268). But the smaller tables—basically triangular though with convex sides—now have an unmistakable period ethos. The designer may have felt that his reason for adopting the shape was purely practical, yet now it seems eloquent of the taste of its time.

Many American bar-cars were much more ambitious. They were part of the effort to give the railroad a bit of glamour. Loewy, in addition to designing locomotives for the Pennsylvania Railroad, provided some opulent interior designs for cars of this type. The basic aim seems to have been to make the passenger forget that he was on a train at all (fig. 269). In one of these cars the windows have been replaced with mirror-glass, which doubles the apparent space and corrects the proportions, but at the same time obliterates the landscape passing outside. The style is the restrained Art Deco of the late 30s, used wherever designers wanted to combine modernity with a feeling of luxury.

A bar-car by the firm of Holabird and Root for the Chicago, Burlington and Quincy Railroad is stylistically more naïve, and therefore more directly revelatory about the designers' aims (fig. 270). They have been less successful than Loewy in getting rid of the long, narrow proportions of the basic structure, but have crammed the design with symbols and reminders of different sorts. Shapes and materials—the use of padded vinyl for the bar surround is a case in point—are reminiscent of the more glamorous kind of Hollywood film set, and are perhaps intended to remove the passenger to an unreal world a good deal more exciting than the one he or she normally inhabited. A particularly telling detail is the use of

266   Interior of an American Pullman car, 1876.

267   De luxe coach for the New York, New Haven and Hartford Railway, 1949.

porthole windows, which seems meant to align this railroad car, in the customer's mind, with the great ocean liners which remained symbols of romance and luxury even in the late 40s.

In fact the 40s seem to have been a rich decade for symbolic design on the railroad. A once notorious example is the weird mock-Tudor buffet-car introduced by British Rail in 1949 (fig. 271). The setting shows a determined effort to evoke an atmosphere of traditional conviviality, though the notion of a half-timbered inn on wheels is worthy of Salvador Dali. By contrast, the railed metal tables are almost aggressively practical. The car caused such an outcry among partisans of 'good taste' that it was soon taken out of service, but a sneaking suspicion remains that it would

268 Restaurant car for the Pere Marquette Railway, 1940s.

270 Bar lounge car for the Chicago, Burlington and Quincy Railroad, by Holabird and Root, 1940s.

269 Bar lounge car in 'The General', designed by Raymond Loewy for the Pennsylvania Railroad, 1940s.

271 Tudor-type buffet car for British Rail, 1949.

have functioned very successfully.

When we speak of good or bad taste in designs like these we are in fact only giving some idea of our own notion of what is appropriate or inappropriate to a particular situation, and this in turn is extremely dependent on our cultural conditioning. Successful design in a railroad bar-car rests on two things—the first is convenience in use, the second—equally important—is the feeling of conviviality it gives, and the fact that it makes the passenger choose the railroad in preference to some rival mode of transportation.

Of all forms of transportation design, that for ships is thought to have the closest natural connection with functional thinking. This is largely based on a misunderstanding. Corin Hughes-Stanton, author of a

brief but highly intelligent study of the subject, long ago pointed out that the supposed historical connection is at least partly a myth: 'The idea that ship interiors should be designed in a "functional" shiplike manner is based on a romantic notion of what cabins and public rooms were like on board nineteenth-century ships.' This assertion is fully borne out by surviving illustrations showing the interior of the *Great Eastern* (fig. 276), even if one may suppose that, as with the luxury liners which followed, her sleeping cabins were more shiplike and practical.

It is significant that, in the seventeenth century, designers of ships were called 'sea-architects', as in fact the designer of ship interiors is very much like the interior decorator who is called in after the architect

273 D-deck cabin on the P & O liner *Orion*, 1935.

274 Cabin on the P & O liner *Oronsay*, 1951.

275 First-class single-berth cabin on the P & O liner *Oriana*, 1961.

272 (*left*) First-class drawing-room on the *Orion*, 1935.

276 A family saloon on the S.S. *Great Eastern, c.* 1858.
277 (*right*) Restaurant area in the Royal Hotel, Copenhagen, with all furniture and furnishings designed by Arne Jacobsen, 1960–1.

has done his best or worst with a building. The spaces may be a trifle more constricted or more awkward in shape, and nowadays safety regulations may limit the range of materials, but these differences are really trifling. Ship interiors respond to the general climate of taste. Thus Edwardian luxury liners like the *Titanic* abandoned the unambitious but cosy bourgeois style of the *Great Eastern* for something more grandiose, which resembled the luxury hotels such as the Ritz which had newly come into vogue on dry land. The public spaces of such liners were a frank statement of the idea that passengers—and most of all those who were travelling first-class—could expect every comfort afloat which they would find in a good hotel on shore. When ship interiors began to make functionalist statements (of a modified kind) once

again, it was not through necessity, but because functionalism itself was in fashion.

In the 30s and subsequently the British company P & O, under the leadership of its chairman Colin Anderson, made a sustained attempt to rethink the design problems presented by the interior spaces of large ocean liners. A sequence of three cabins from three different P & O ships—the *Orion* of 1935, the *Oronsay* of 1951, and the *Oriana* of 1961 figs. 273–75) —shows that the company's designers soon arrived at a formula which varied remarkably little in the course of 26 years. It is largely portable objects—a Cubist rug or a chair which is a moulded shell on metal legs— which give away the precise date. Comfort is adroitly fitted into a small space, and an air of luxury added with fine woods and cabinet-work.

The public rooms of these ships were a different story. The most notable was the earliest in the sequence, the R.M.S. *Orion*. Her interiors, and those of her successor the *Oronsay*, were designed by Brian O'Rorke. He, like Loewy working for the Pennsylvania Railroad, used a restrained version of the already well-established 'Deco' style (fig. 272), and clearly for similar reasons. The *Orion's* public rooms were a statement, not about the sea as such, but about the get-away-from-it-all luxury of life on board ship.

Design histories often take liners, and also hotels, as infallible indicators of the evolution of decorative style. This is true only up to a certain point. A hotel, like a liner, can become a rigidly controlled environment which in design terms has an exemplary function. It shows the public what the possibilities are at a particular moment. The *Orion* is a good example for 1935, and Arne Jacobsen's Royal Hotel in Copenhagen is an even better example for 1961. Because such interiors offer the public only temporary shelter—the assumption is that everyone there is a transient—they can afford to be both dictatorial and idiosyncratic (fig. 277). As either passengers or hotel guests, people accept things they would not want to live with every day.

One important point about such commissions is not merely that they serve to publicize a particular attitude to design or even a specific style, but that they serve as test-beds for new ideas. The elegant steamer-chairs specially designed for P & O by Ernest Race are an authoritative restatement of the possibilities offered by moulded plywood. Often items specially designed become the prototypes for goods which are later put on sale commercially (fig. 278). But it is only rarely that designers can exert total control over an environment, however much they would like to do so. Design is more usually the result of an encounter between the designer, the manufacturer and the customer in circumstances where the boundaries are much less rigidly defined.

278 Holloware designed for the *Oriana*, by Robert Welch, 1960.

279 (*opposite*) Desk, the top veneered in mulberry, and the rest bound in buffalo suede, by John Makepeace, and made in his workshops at Parnham House, Dorset, 1978.

The Fashionable Interior

Stylistic change shows itself most clearly at the fashionable end of the market. It also manifests itself with greater visibility in objects which are meant from the first to have some decorative function, as well as a purely useful one. The typical decorative idiom of the 30s leaps to the eye when one looks at a roller-printed fabric designed by Irene Fawkes in 1935. This, with its borrowings from Cubism (fig. 280), summarizes one aspect of the taste of the period.

Non-domestic objects are, of course, well able to evoke a particular epoch—a Swedish kiosk and street litter bin of 1948 (figs. 281 and 282) are already redolent of the conscious, rather ersatz jollity which overtook European design in the years immediately following World War II. But interiors are even more evocative—and it is in this context that we tend to remember designers' names, just as we remember the names of fashionable couturiers when trying to summon up the atmosphere of a particular time. The reader can test the truth of this assertion against, for instance, a London coffee-bar designed by Terence Conran in 1956 (fig. 283), or two interiors by the celebrated Italian designer Gio Ponti, created for the Casino at San Remo in 1952 (fig. 284). It is interesting to note that the two designers, though they came from very different backgrounds, also seem to have

something in common. There is a liking for ornamental motifs which are simplified but self-consciously fanciful, scattered seemingly at random. The furniture used in these interiors is equipped with the spiky legs which now seem typical of the epoch.

Neither a coffee-bar nor a casino can be described as a domestic interior in the full sense of the term. Nevertheless, professional work of this type gives a lead to the ordinary domestic setting—it provides the consumer with something to copy and adapt. It also pinpoints the nature of a particular style by exaggerating its characteristics. In 1957 Gio Ponti designed a room setting for the Milan Triennale (fig. 286). Of all the objects chosen, only the Charivari dining chair on the extreme left of the picture seems in the least 'contemporary' today. It is perhaps no coincidence that this is the only piece of furniture included which is based on a traditional design—a version of this chair has been manufactured in Italy since the early nineteenth century. Even more firmly

located in its period, when we look at it now, is a room setting from the immediately previous Triennale of 1954 (fig. 285). This, with its diagonals, triangles and irregular quadilaterals, is very nearly as divorced from today's ideas about interior design as an interior by William Kent or Robert Adam.

It can be argued that it is objects combined in this way which immediately summon up a period—that domestic items seen in isolation are a good deal more neutral. In fact, a number of distinctions have to be made. It may be the decoration, rather than the form of a thing, which ties it to a particular epoch. In 1933–4 the firm of A. J. Wilkinson of Burslem commissioned a group of contemporary English artists to produce designs for decorating chinaware. Among them were Graham Sutherland, John Armstrong, Dod Proctor

280 (*left*) Roller-printed fabric by Irene Fawkes, 1935. London, Victoria and Albert Museum.

281 (*right*) Rubbish bin at Lake Malaren, Stockholm, designed by Stockholm Parks Department under Helge Blom, 1948.

282 (*above right*) Swedish street kiosk at Karlskoga-Bofors, 1948.

283 *The Orrery*, coffee-bar designed by Terence Conran, 1955. From *Architecture and Building,* March 1955.

284 Interior at the San Remo Casino, designed by Gio Ponti, 1952.

285 Room designed by Gregotti, Meneghetti and Stoppino for the Milan Triennale, 1954.

286 Apartment interior designed by Gio Ponti for the Milan Triennale, 1957.

287 (*opposite*) Commemorative mug designed for Wedgwood by Eric Ravilious, 1940.

288 *Chevaux*, chinaware
designed by John Armstrong
for A. J. Wilkinson Ltd.,
Burslem, 1934.

289 *Marine* design by Dod
Proctor, Bizarre ware by
Clarice Cliff, made by A. J.
Wilkinson Ltd., Burslem, 1934.

290 *Circus* design by Laura
Knight, Bizarre ware by Clarice
Cliff, made by A. J. Wilkinson
Ltd., Burslem, 1934. London,
Victoria and Albert Museum.

291 *Universe* design by
Graham Sutherland, Bizarre
ware by Clarice Cliff, made by
A. J. Wilkinson Ltd., Burslem,
1933–4.

and Laura Knight (figs. 288–91). The paintings have a folksiness typical of the time, and one can trace the influence of the Bloomsbury-inspired Omega Workshops of more than a decade earlier. Sutherland was also commissioned to make designs by another firm, E. Brain & Co., at precisely the same epoch. His 'White Rose' pattern gives a fascinating insight into another of the decorative preoccupations of the time—the Victorian revival, which had been gathering strength for a decade (fig. 292). Similar ideas, rather more lightly expressed, are found in Eric Ravilious's commemorative mugs—one for the coronation of King George VI, and the other for the centenary in 1940 of the foundation of the new Wedgwood factory in Barlaston (fig. 287).

But the forms to which these decorations are applied are virtually neutral, taken from stock, and the Wedgwood mugs in particular are a pattern which had been current since the eighteenth century. We can talk of design as such, rather than just decoration, when the form as well as the pattern changes. China from Ridgway & Addersly, exhibited in a show of

'Contemporary Tableware' in 1954, shows scattered linear patterns typical of the decade applied to shapes which now seem in some cases to be very much of the period—for example, the dish-shaped plates without rims (fig. 293). A 'Homemaker' plate of the same epoch (fig. 295) takes the same form, and in this instance the scattered motifs are typical examples of 50s furniture—note particularly the boomerang-shaped coffee table in the centre—so that the result now seems doubly eloquent of its time. But having noted this, it is still impossible to guess why the rimless plate suddenly gained favour.

The juxtaposition of certain domestic objects which belong in more or less the same category can bring home to one quite forcefully the way in which preferences evolve, seemingly without rational motivation. A 50s glass vase of folded handkerchief shape (fig. 296) shows a free and informal treatment of the material which is very different from the Bauhaus-inspired formality of a heat-proof glass tea-kettle designed by Signe Persson for the Swedish firm of Kosta-Boda in 1970 (fig. 294). There is a big difference

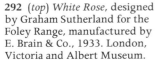

292 (*top*) *White Rose*, designed by Graham Sutherland for the Foley Range, manufactured by E. Brain & Co., 1933. London, Victoria and Albert Museum.

293 (*above*) Tableware by Ridgway & Addersly Ltd., decorated by Peter Cave, 1954.

294 (*right*) Kettle in heat-resistant glass designed by Signe Persson for Kosta-Boda, 1970.

295 (*above right*) *Homemaker* plate, *c.* 1952.

296 (*above far right*) Glass Handkerchief vase by Chance Brothers, 1950s.

in function between the two objects, but even this does not quite explain the huge difference in approach.

Technological changes do, however, affect the development of household furnishing, even at the top end of the market, and so do economic and theoretical considerations. In Britain, theory was given its head in the immediately post-war epoch which saw the tail-end of an attempt by the 'great and good' to impose their own standards on the mass market. In this they were aided by the shortages of the time. These dictated that the supply of new furniture must be under strict control, and that it must be made of plywood—the cheapest and most easily available material. The various 'Utility' ranges, first manufactured during the war, continued into the late 40s (fig. 199). The rigorously simple forms were post-Arts and Crafts, reminiscent of the furniture designed in the Edwardian epoch by Ambrose Heal. But public taste was not to be changed under duress, though it may be coincidental that 50s design often has a guilty look—a wish for opulence in conflict with an essential spikiness, an emphasis on fun and freedom which is nevertheless rather shamefaced. Even the puritans who created the Utility range were not immune to the sensibility of their time. The roughly triangular 'free form' of the occasional table shown to the right of the Utility group shows a movement from 30s Streamline towards what we now regard as the typical furniture shapes of the 50s.

In America a similar movement was taking place. A boomerang-shaped 'buffet-table desk' with tapering legs, designed by Edith and William Fernandez, incorporates many of the favourite notions of the day. It was sufficiently in tune with the times to be awarded a Certificate of Merit in a design competition held in 1948 (fig. 297).

The extensive use of plywood in Utility furniture did, however, herald something important—it signalled a shift away from natural materials. Thirty years later a handmade desk by John Makepeace (fig. 279) would seem special and luxurious not only because it was a one-off by a noted craftsman, but because it was not made of plastic and plywood.

298 Lucite bed designed for Helena Rubinstein by Landislas Medgyes, 1920.

One valid way of examining twentieth-century furniture—even that avowedly made for the luxurious end of the market—is through the relationship of style, and the obsession with style, to the designers' approach to new materials. I am speaking here not only about the steel tubing beloved of the Bauhaus designers (as much for symbolic as for practical reasons), but about plastics of all kinds.

Plastics have had a symbolic function in twentieth-century design—as a sign of committment to modernity and a new age—in addition to purely practical ones. It was perhaps natural that symbolic values attracted designers even before they began to try to discover the real possibilities offered by this new range of materials. They began by treating plastic as a more tractable kind of glass or even hardstone. One of the earliest instances of its use in any ambitious way for furniture seems to have been the lucite bed made for Helena Rubinstein in 1920 (fig. 298). This bed was the ancestor of a whole range of 'glamour' furniture which was still being made under Holly-

297 Buffet-table desk in natural birch, by Edith and William Hernandez, New York, 1948.

299 Rainbow coffee-table in laminated acrylic, by John Williams, 1970s.

wood influence some 30 years later (fig. 300). This furniture stuck very firmly to traditional forms—chiefly to a bastard version of French Empire. It also stuck, as far as the material allowed, to traditional construction. At least part of the frisson it imparted came from the clash between form and material, which reflected a cultural clash in *nouveau riche* society which was avid for what was new, yet still anxious to ally itself with what was established.

Even today craft furniture-makers are occasionally attracted to acrylics because of their transparency, and the possibilities they offer for glowing colour. One recent result of this attraction is the laboriously hand-made coffee table illustrated (fig. 299). It contains over 100 multicoloured plastic sections, glued, machined and polished.

But plastics also offer the possibility of constructing furniture in completely new ways, and it is here that the designer enters the picture in a much more fundamental sense, since the new forms he devises are not dependent on stylistic choices alone, but on the actual method of manufacture. It is, of course, true that this quest can go wrong—the result being a rather sterile kind of technological novelty. For me, this sterility is typified by the blow-up plastic furniture which was briefly fashionable in the late 60s (fig. 301). Blow-up furniture was predicated on a way of life where everything would be fast-moving and ephemeral—where it didn't matter if sofas had little more durability than soap-bubbles, and where it seemed appropriate that they should rival soap-bubbles in lightness and transparency. These ideas were defeated first by a lack of comfort in the furniture itself, and secondly by an ineradicable human desire for a reasonably stable and reassuring domestic environment.

300 American 'glamour' furniture in clear plastic, in a room setting with mirror-glass walls, 1950.

Far more significant is the marriage of taste and technology to produce something new represented by a famous Saarinen chair of 1957 (fig. 195). This is genuinely revolutionary both in appearance and in method of manufacture. The design was the culmination of experiments which had been going on for nearly two decades. It consists of a reinforced fibreglass shell poised on a painted aluminium base. The two elements seem to flow into one another, but are in reality quite separate. Eero Saarinen and Charles Eames started to experiment with moulded polyester in 1940, in response to a competition for 'Organic Design in Home Furnishings' backed by Bloomingdale's, the New York departmental store. Eames won it, and his entry and others were exhibited the

following year at the Museum of Modern Art. The competition led not only to a new approach towards the use of plastics in furniture (development was speeded by the demands of the aircraft industry during the war), but to a new way of thinking about seat-furniture, where the seat unit was quite separate from the supporting structure.

In fact, this was an example of something significant in design—a solution devised in response to the possibilities and limitations of a particular material, often something novel, is then discovered to have a more general application.

Implementation of the idea that seat and supporting structure are two different things gives a common look to many chairs designed in the 50s, no matter who

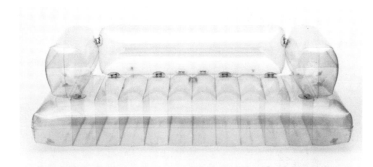

301 Blow sofa designed by De Pas, D'Urbino and Scolari, and made by Zanotta of Milan, 1967.

302 .Upholstered chair, Indian laurel legs with orange tweed upholstery, designed by Neville Ward for Scottish Furniture Manufacturers, 1950.

the designer is, or what the materials chosen. One can see this approach at work in Ernest Race's 'Antelope' stacking chair in resin-bonded plywood and metal, designed for the Festival of Britain (fig. 303), and in Arne Jacobsen's similar stacking chair '3107' (fig. 304), though here the legs are bolted separately to the seat, so the structure has been to some extent elided. Yet another version of the same design philosophy is represented by Nigel Walters's 'Lacewing' chair of 1953 (fig. 305), with a frame made of steel rods and a seat of metal mesh.

Oddly enough, the general 'look' of these chairs, which is so dependent on their actual construction, was occasionally transferred to designs where the construction remains conventional (fig. 302)—the influence of technology upon style and style upon technology can thus be thought of as reciprocal.

Plastics did not simply suggest a new way of making chairs—they became basic to the whole furniture industry, and have been used increasingly in upholstered furniture. Upholstery made with horse-hair and coiled springs—itself an early nineteenth-century innovation—began to be driven out by plastic granules and plastic foam, which were far less labour-intensive at a time when labour costs were rising steeply. Designers became increasingly accustomed to the new plastics, and increasingly free and confident in the way they handled them—a development prompted not merely by the materials themselves but by what was going on in the fine arts. More and more seat-furniture became sculpture for the domestic environment, an equivalent of the abstract forms young sculptors were producing at the same time. One of the most striking examples is the supremely elegant 'Djinn' chaise-longue designed by Olivier Mourgue in 1963, at a time when the fashion for large-scale, brightly coloured sculptures made of fibreglass was at its height (fig. 307). The 'Dijnn' became almost as much a symbol for its own decade as the Le Corbusier chaise-longue did for the 20s (fig. 306). Mourgue has this much in common with Le Corbusier—he uses a tubular steel frame. But it is completely hidden by polyether foam which in turn is encased in a removable cover of stretch-nylon jersey. Upholstery and covering are thus both products of the plastics laboratory.

The 'Dijnn' is not the only piece of contemporary seat-furniture which aspires to the condition of sculpture. Modern chairs tend to have especially forceful and individual shapes (figs. 308 and 310) because they give visual articulation to spaces which are increasingly plain and undifferentiated—qualities due to the contemporary preference for built-in furniture.

Whereas one range of problems confronting today's furniture designers is new, another range has a long history, already touched on in this book. Like the late eighteenth-century cabinet-makers, modern designers produce pieces which have multiple functions to alleviate a chronic shortage of living space. Their solutions are often just as ingenious and elegant as those invented by men like Sheraton and Hepplewhite. Typical is Franco Poli's table for the Italian manufacturer Bernini (fig. 309), which is completely adjustable in height and can serve either as a coffee table or a dining table. But this table does show a significant deviation from eighteenth-century design philosophy. The mechanism is boldly emphasized, where an earlier cabinet-maker would probably have compromised its efficiency by trying to conceal it. Though the basic device is very simple—a

303 (*extreme left*) Antelope chair designed by Ernest Race for the Festival of Britain, 1951.

304 (*far left*) Stacking chair '3107', designed by Arne Jacobsen, 1957. London, Victoria and Albert Museum.

305 (*left*) Lacewing chair in metal rod and mesh, designed by Nigel Walters, 1953.

306 (*above*) Le Corbusier, with Pierre Jeanneret and Charlotte Perriand, chaise-longue, 1928.

307 (*right*) Djinn chaise-longue designed by Olivier Mourgue for Airborne, Paris, 1963.

308 (*opposite*) *Fiora* chair by Gigi Sabadin, 1970s.

309 (*left*) Variable height table by F. Poli for G. B. Bernini, 1970s.

310 (*above*) Circle chair in mahogany and natural coach hide, by Jorgen Hovelskov, 1970s.

crank which either spreads out the support or brings its component parts closer together, thus either lowering or raising the table-top—the designer's assumption that we will find the means whereby this is done fascinating makes him very much a man of our time.

One can appreciate the intricacy of the relationship between technological imperatives and merely fashionable ones in top-class domestic design only if one also grasps the underlying premise that technology is itself something which fills us with emotion, which simultaneously seduces and repels. Home furnishings offer not merely solutions to practical problems, but props for the stage-sets on which we act out our lives. Because of this one must never underestimate the importance of their emotional and symbolic content. This, the most discussed area of design, the one closest to the general public consciousness, is also the one where it is hardest to discover principles which can be generally applied.

Machines in the House

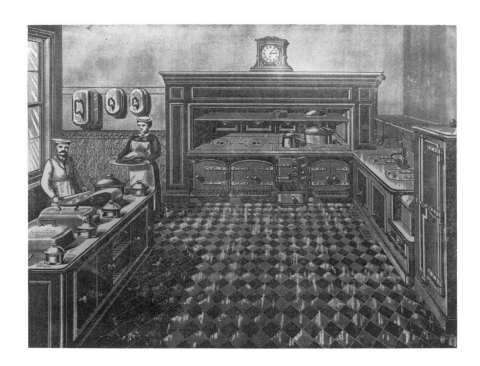

Industrial design entered the domestic environment through the back door which led to the kitchen. This was logical, since a kitchen could be thought of as a factory operating on a very small scale—a factory for producing meals, transforming raw materials of various kinds into a finished product. The necessity for rational kitchen planning was understood quite early. An illustration from a British catalogue of around 1890 (fig. 311) shows that today's fitted kitchen has a long pedigree. The manufacturer, William Froy & Sons, states in the accompanying text that 'if clients will kindly send us dimensions of kitchen and requirements, we will send a complete estimate with drawings'. This late nineteenth-century kitchen can be compared in many respects with a fitted kitchen of 1946 (fig. 312), itself the predecessor of the admittedly more elaborate off-the-peg kitchens we know today.

However, during the period between the two illustrations an important social change had taken place. This was the disappearance of the servant class. Even in the 1870s not much more than 20 per cent of the British population had a living-in servant. The proportion declined rapidly as the middle class expanded, and as servants found other and pleasanter occupations in factories. In the kitchen of 1946, the housewife is shown at work herself, without help, and it is a brighter and more cheerful place. It is also a place where domestic appliances were more and more welcome, not merely because they lightened the work-load, but because they gave a certain prestige to what had previously been regarded as menial.

Even before this, human ingenuity had been busy devising ways of lightening domestic labour. In the Middle Ages spits for roasting were turned by dogs confined in cages. Later people tried weights of the kind used to run a clock for the same purpose, or vanes which turned in the draught of hot air rising through the kitchen chimney. An interesting, though fairly late, example of one of these devices is a nineteenth-century reflecting oven with clock jack to turn a joint of meat (fig. 313). Intended to stand before an open fire, this is simple but also ingenious—a good application of design logic. The reflecting oven additionally demonstrates how, long before the coming of gas and electricity, designers were devoting thought to the conservation of energy, and looking for the most efficient way of utilizing the heat source available to them. A solid-fuel oven of *c*. 1860 is divided into several compartments according to function. It includes a plate-warmer, a hob, and no fewer than four ovens (fig. 314).

Though solid-fuel stoves like the Aga still enjoy a market today, the real challenge to designers of cookers in the late nineteenth century lay in finding the best way of using gas. Later still, gas was challenged by electricity. A late nineteenth-century gas stove (fig. 315) is recognizably a more primitive version of the kind of stove which is still in use today. The same is true of a Belling electric stove manufactured in 1919 (fig. 316). The most prophetic design, however, is the Archer system electric cooking outfit, which appeared in the catalogue of the British General Electric Company as early as 1909 (fig. 319). This has a

311 (*opposite*) Fitted kitchen, illustration from a catalogue issued by William N. Froy & Sons, *c* 1890.

312 (*above*) Fitted kitchen, 1946.

313 (*left*) Nineteenth-century reflecting oven with clockwork jack, Dearborn, Michigan, collections of Greenfield Village and the Henry Ford Museum.

314 Solid fuel range produced by Newton Chambers & Co., of Thorncliffe near Sheffield, *c.* 1860.

315 Nineteenth-century gas cooker.

316 The Modernette—the first complete Belling electric cooker, 1919.

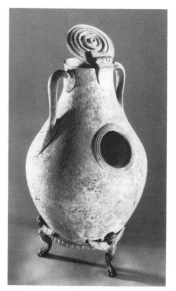

317 Roman water heater, AD 79. London, Science Museum.

318 Wilson's gas kettle, 1888. London, Science Museum.

319 Archer system electric cooking outfit, 1909.

number of appliances—a skillet, a saucepan and a steamer—attached by leads to the main unit. Later these appliances were to be hived off, and would become the array of small-scale kitchen devices we know now.

In fact, one of the industrial designer's major opportunities really came with a plethora of domestic gadgets, the vast majority powered by electricity. The idea of small appliances of this type was not new. In a certain sense the Romans had thought of them, as is proved by the graceful bronze water heater of the first century AD, built to take a handful of glowing charcoal in its belly (fig. 317). The Victorians also hankered after such portable appliances, and rather clumsily tried to use gas for the purpose (fig. 318). But it was electricity which really opened up a full range of possibilities.

A typical modern electric kitchen appliance is the blender, now installed in almost every kitchen. A blender makes use of simple rotary motion to mix ingredients together, and also in some cases to mince or shred them. Early models, like the Westinghouse

Electric Food Mixer (first produced in 1947), the Kenwood Chef (1950), and the GEC version (1946) are conceived in very much the same anthropomorphic terms (fig. 320–2) where a motor replaces the hand of the cook. A set of rotary beaters is placed slightly off-centre over a mixing bowl. Sometimes the mixer can be detached from its base and used as a portable unit, and sometimes different attachments can be used, for beating, whisking or kneading, and slicing or shredding in addition. The Westinghouse model in particular is a fine example of the American Streamline Style of the period.

The electric blender developed in two different ways. The first was stylistically. In later but similar models, for example, the arm which holds the beaters over the bowl is squared off (fig. 323). In a minor key, blenders thus followed the lead being given by Detroit auto styling, and styles seem to have changed for the same reason—in order to make previous appliances seem obsolete. But in some models a different principle was introduced—the blades now rotated within the container, and the motor was housed in the base of the

320 Westinghouse electric food mixer, 1947.

321 Kenwood Electric Chef, 1950.

322 GEC electric mixer, 1946.

323 Kenwood Major, model A 907D, 1981.

324 (*extreme left*) Kenwood blender de luxe.

325 (*far left*) Kenwood Major A 707A mixer and blender.

326 (*left*) Kenwood Food Processor de luxe, model A 530C.

machine (fig. 324). Appliances of this type have a slightly different function—they shred vegetables and fruit more efficiently than they beat eggs or cake mixtures. Some manufacturers, Kenwood among them, acknowledged this fact by creating more elaborate versions which combined the new blender with the older blender-beater (fig. 325). Essentially, however, it was the blender with a motor in the base, now rechristened a 'Food Processor' and equipped with a variety of rotating disks for shredding and slicing, as well as blades for mixing and cutting, which seemed likely to triumph (fig. 326).

The food processor was in its own way an even more significant novelty than the blender. The blender implied the intensification of the wish to escape the burden of kitchen drudgery. The food processor was eloquent about changed dietary tastes and increasing culinary sophistication. Not only was it almost essential for the preparation of the shredded raw vegetables and the fresh juices increasingly recommended by dieticians, but it also put within reach of those with little time at their disposal the mousses, pâtés and other delicacies which were formerly only made by

extremely patient and expert cooks.

Electricity was never confined to the kitchen. Its cleanliness and controllability made it welcome throughout the house. Peter Behrens of AEG was the first of a long line of industrial designers called upon to find appropriate and convenient forms for a wide range of non-culinary electric appliances. In many of these the desire to find a form which was expressive from the symbolic point of view struggled with considerations of appropriateness and practicality. A certain amount of mockery has been directed by design historians at the idea of an electric fan in Streamline Style (fig. 327), since a fan is essentially a static object which circulates the air but does not move through it. The metaphor in the designer's mind was clearly that of an aircraft engine in its nacelle, and I find that in some strange way it does help to make sense of the object. Clearly, the stylistic device, however irrational, communicated to consumers, since it was widely used.

Electric fires, on the other hand, have an unhappy design history which is explained but not excused by their symbolic function as a source of warmth which replaces the old-fashioned fireplace. An early Ferranti model uses the heat source very directly—it is an element whose warmth is enhanced and directed by a curved reflector (fig. 329). But the manufacturers, and perhaps the public too, could not long remain content with such unadorned simplicity. The element and its reflector were put in a metal frame with vaguely Georgian connotations. More astonishingly, they were even framed between a pair of traditional andirons (fig. 330). It was not until the 50s that the bar fire had become sufficiently commonplace to return to its original simplicity, though by now it was also provided with a protective metal grid (fig. 328).

Yet people continued to feel that the electric bar fire was somehow inadequate as a symbol, no matter how efficient it might be as an actual means of heating. The result was the coal-electric fire whose flickering red glow imitated the comforts of an open hearth. These too have been much mocked by design historians and other guardians of good taste, but the public has bought them in formidable quantities despite the frowns.

Electric fires, like electric fans, raise not merely the question of design which is either successful or unsuccessful within a given formula, but the question of whether the problem might not be better solved by choosing a different solution altogether—i.e. by installing central heating or central heating combined with air-conditioning. The growing tendency to opt for the 'invisible' solution is part of a general trend since World War II to dematerialize consumer products—something which occurs when a single product is replaced by a system which may be either concealed or dispersed (like a hi-fi set up) so as to become inconspicuous.

Experiments have certainly been made with centralized dust-extraction systems in modern build-ings, but the vacuum cleaner, like the electric blender, remains one of the most typical domestic consumer products. It has in fact a much longer history than the blender. At first there were hand-operated cleaners (which needed two people to operate them) as well as electric ones, but even the electric models go back to the early years of the present century. Some early models, like the B.V.C Model 'R' of 1910 (fig. 331) are

327 Streamlined table-fan by Limit Engineering Group Ltd., 1956.

distinctly industrial looking. By contrast, the original Hoover (fig. 332), though it dates from three years earlier, is comparatively tidy in shape, and the decoration coyly hints that the machine is meant to be used by the lady of the house. This is not the only interesting point of difference—the two machines offer the two basic solutions we still see today—one where the cleaning head is attached to the motor by a long flexible hose, and one where the head and motor are integrated into one unit. In fact, by 1926 the vacuum cleaner had achieved not one definitive shape, but a quasi-definitive pair of alternative shapes, and these have changed very little—as a series of illustrations of one type, dating from the mid-20s to the late 40s (figs. 333–4) will demonstrate. The one significant alteration in this design was to move the motor from a vertical to a horizontal position, to give a

329   Ferranti radiant electric fire, model no. 1, 1927.

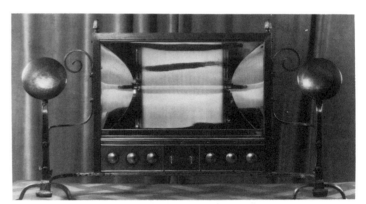

330   Ferranti radiant panel fire with wrought iron fire-dogs, 1920s.

328   Portable Ekco electric fire, designed by Joseph Carr, 1954.

sleeker shape. One can hardly describe this as a fundamental technological change.

More recently, the vacuum cleaner appears to have suffered from a kind of technological overkill (fig. 335). Many refinements have been added in the latest models—variable power, a remote control device, a wider range of interchangeable cleaning heads. The basic concept still remains precisely the same—a motor to create a vacuum which sucks up dust and deposits it in a bag or other receptacle. One wonders how many housewives, seduced into buying the Hoover Sensotronic by skilful advertising, still

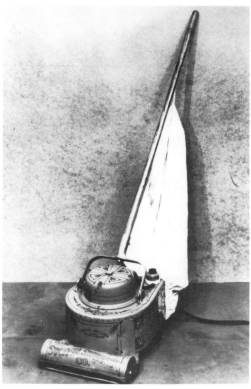

332 Hoover electric vacuum cleaner, 1907.

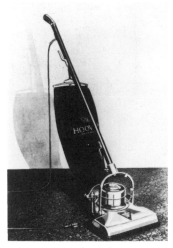

331 (*top*) Model 'R' electric vacuum cleaner by B.V.C., 1910.

333 (*above*) Hoover model 700 vacuum cleaner, 1926.

334 (*right*) Morphy Richards vacuum cleaner, model V.C. 18, 1949.

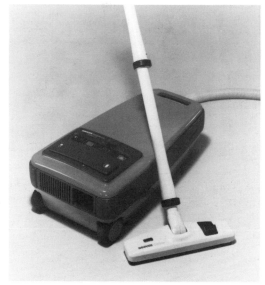

335 Hoover Sensotronic system 2, model S3 128, 1981.

336 Videomaster 24-tune door chimes, designed by Stephen Frazer, 1977.

337 (*right*) Richard Hamilton, *Toaster*, 1967. Metallized print, 35 × 25 in. (89 × 63.5 cm.). London, Tate Gallery.

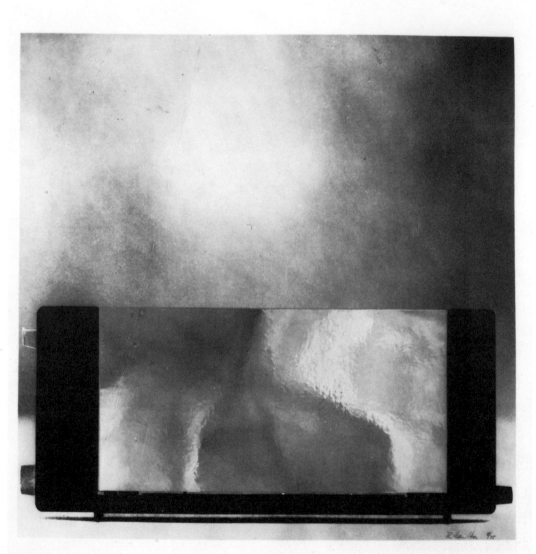

Toaster

New, practical, outstanding, this print was made possible by a number of fresh ideas. The proof of the excellence of the toaster that inspired this work of art has been supplied by the results of severe endurance tests recently performed. The appliance was kept working for a total of 1458.3 hours (not counting brief periods for cooling). This was the time taken to toast 50 000 slices of bread. That is a pile of bread well over a quarter of a mile high.
Just how outstanding the design is can be proved by the fact that it has been

included among the most attractive objects for everyday use exhibited at the New York Museum of Modern Art – the only automatic toaster in the world to achieve this honour.
White bread, black bread or even rye bread? Ask your friends and neighbours and they will tell you that toast is a first-class delicacy. It tastes good and has never been the cause of anyone losing their driving licence. It keeps you fit and your body needs it.

Printed on Saunders plain mould special printing s/o demi 80.5 lb/500 (complete with Marlerfilm and Marlerflex ink and applied metallized silver polyester) in an edition of 75.
Dimensions 25" wide, 35" high, image area 23" square.

hardly bother to use many of the facilities it offers.

Some ultramodern domestic machines, approved first by the design community, and later by consumers in general, have been elevated to the status of household idols, which confer mysterious prestige by their very presence. This tendency is cleverly satirized by Richard Hamilton in his 1967 print featuring the famous Braun toaster (fig. 337). His version, subtly altered by hand, also includes Braun's boast that the toaster had been 'included among the most attractive objects for everyday use exhibited at the New York Museum of Modern Art—the only automatic toaster in the world to achieve this honour'.

Objects like the Braun toaster lead one to reflect on some of the niceties of functional theory. It is an object which, in addition to looking good, undoubtedly makes toast and makes it well. One nevertheless wonders if the way it looks isn't perhaps more important to the purchaser than the functional purpose—and the symbolic function, the prestige ownership confers, is undoubtedly important too. Appliances have become more and more specialized, just as furniture did in the late eighteenth century. Most middle-class households possess an array of them, each designed to perform a specific task. Taking them out of the cupboard and setting them up, then cleaning them afterwards may in fact involve more effort than doing the same job in an old-fashioned way by hand. Or else there are appliances which, so to speak, invent tasks and then proceed to carry them out. An extreme example—and another instance of so-called 'black box' design—is a product called the Videomaster Door Tunes (fig. 336). This is a replacement for the door-knocker or doorbell—a device which can be set to play anything from 'Twinkle, Twinkle Little Star' to a few bars of Bach's Toccata in D Minor. This is clearly an object of considerable technical ingenuity, and the casing is elegant. But to what extent can we think of it as doing a neccessary job in the household? It is in fact no more than a toy. And once a supposedly functional device reveals itself to be really no more than a plaything, our concepts of 'function' and 'efficiency' are inevitably revealed as rather shallow.

Technological Toys

To call the items discussed in this chapter and the next toys is no doubt provocative. All of them are meant to have some useful function, though the function often seems secondary to other considerations. The vast majority are things which are not absolute necessities, and sometimes only marginal in terms of making daily living easier or more convenient. Many are things with a strong role in conferring status. Some are involved with the miniaturization which is part of the fascination of many toys, and some are things specifically designed to entertain or otherwise help pass the time.

Clocks and watches do not pass the time—they record it passing. But with them the prestigious connotation is often very strong, and it is linked to the idea of possessing and controlling technology superior to that of the next man. Clocks were perhaps the first complex machines to become an accepted and commonplace part of home furnishing; and watches, which are miniature clocks, were certainly the first truly portable mechanisms of this degree of complication, designed to accompany the owner wherever he went. Their prestigious connotations were strongly reinforced by their inevitable connection with dress and fashion. Clocks similarly responded to changes in architectural style. If one compares a bracket clock of *c.*1695 with a Neoclassical mantelpiece clock by Vulliamy of about 1800 and a spiky Gibson electric wall clock of 1955 (figs. 339–41), one is confronted with a panorama of decorative styles. The earliest clock of the three, which is by the famous English clocksmith Thomas Tompion, is already highly developed from a technological point of view. There is a small dial at the top left of the clockface for regulating the pendulum, while the striking mechanism can be silenced by turning the hand on the right. The maker has provided a mock-pendulum in a curved opening at the centre of the main dial, which moves with the real pendulum (which is concealed) to show whether the clock is going or not.

Different as they are in style from one another, these clocks have more in common than any one of them has with a more recent timepiece—a digital alarm clock designed in 1976 (fig. 338). At first sight this doesn't look like a clock at all, but can be compared perhaps to a miniature television set. Any allusion to a traditional clockface has been discarded, and the time is displayed in arabic numerals in a liquid diode display behind a plastic panel. This clock belongs to a whole family of small electronic devices which have recently reached the domestic market, among them a large number of ever-smaller pocket calculators which display their results in the same way, and make use of the same technology based on the microchip (figs. 342 and 346). These small calculators, like the digital alarm clock, are essentially mysterious containers – their inner workings are concealed and do not necessarily conform to the shape of the outer casing. It is enough that the mechanism fits within it. Nor is it immediately apparent to the uninitiated how the object is meant to be used.

It must be admitted that watches may also have possessed some of this mysteriousness when they were first introduced as princely luxuries in the sixteenth century, having evolved from the miniaturization of the drum-shaped table clocks of the time.

338 (*opposite*) Electronic digital alarm clock designed by John Ryan for House of carmen, 1976.

339 (*right*) Table-clock by Thomas Tompion, *c.*1695.

340 (*far right*) Mantel clock by Benjamin Vulliamy, *c.*1800.

341 (*extreme right*) Gibson Sunray electric wall clock, designed 1951.

342 Pocket electronic calculator by Braun AG, designed 1976.

What made them technically possible was the invention of the spring drive. But the watchmaker also had to find ways of regulating the spring, so that its energy was released smoothly and evenly, and the story of the development of the watch is from a horological point of view largely a story of constant ingenious improvements in this respect.

Watches were for long too large, and too sensitive to shock, to be worn on the wrist, as is the usual custom today. They were carried in the pocket. Nor were they necessarily round—an oval shape was common in the mid-seventeenth century (fig. 343). Even when timekeeping was still a somewhat uncertain affair watchmakers sought to introduce complications of various kinds, perhaps to distract attention from the relatively poor performance of the main mechanism. The watch illustrated, made by Benjamin Hill, who was master of the London Clockmakers' Company in 1657, has a dial which indicates astrological and astronomical phenomena as well as one which tells the time, and there are subsidiary indicators for the date, the day of the week and the phases of the moon. In conformity with the taste of the time, the whole watch-face is elaborately engraved.

Seventeenth- and eighteenth-century watchmakers, even when striving for the utmost accuracy, could not usually resist a display of old-fashioned craftsmanly skill, as well as a less obvious one of horological finesse. One spur to the improvement of timekeepers was the fact that it was impossible to navigate accurately without them. This became more and more of a problem as longer and longer voyages were undertaken along the world's trading routes. In 1714, on the advice of Isaac Newton, the British Government offered a prize of £20,000—then a truly immense sum of money—for a maritime chronometer able to ascertain longitude to half a degree. The test period was six weeks, and this meant making a timekeeping instrument with an accrued error of no more than two minutes during the set period—an average gain or loss of about three seconds a day. After three previous attempts John Harrison won the prize in 1759, with what was in fact a very large watch (figs. 344,345). Despite the severity of the technical task the watchmaker had set himself, he could not resist a few flourishes, as the traditional pierced and engraved plates of his No. 4 Chronometer show.

A maker who exercised a huge influence over the development of the modern watch was the Swiss-born Abraham Louis Breguet, who worked in Paris before, during and after the French Revolution. Breguet was responsible for a remarkable list of technical innovations. One of his most important inventions was the tourbillon, which allowed the entire balance and

343 Oval watch made by Benjamin Hill, London 1657. London, British Museum.

344, 345 Harrison's no. 4 chronometer, 1759, front view and movement. London, Science Museum.

escapement of the watch to revolve so as to avoid the position errors which occurred when the mechanism was on its edge. Breguet worked for the very rich, and he was patronized by most of the crowned heads of Europe. His watches appealed equally to the family of Napoleon—Lucien Bonaparte in particular—and to the Bourbons. Even by the standards of his day, his watches were expensive objects. But very commonly, especially if they were watches for men, the design of the case, and of the face and hands, was extremely plain and practical. In this Breguet was following the sartorial fashion of the day, where Beau Brummell and other dandies had brought about a great reform in men's dress, which grew increasingly severe. But the result was watches which were much easier to read, and therefore more practical in use.

Some of Breguet's work was in direct response to fashion—his watches became flatter, so that they could be slipped into the pockets of tightly cut clothes without making an unsightly bulge. But it was other innovations which make Breguet notable for the history of industrial design. It is clear that he located and analysed specific problems just as a modern designer does. To counteract the imperfect lighting of the day, he invented the *montre à tact*, with a raised single hand and a series of studs around the edge of the watch which allowed the hour to be read with the fingertips. Perhaps more significant still, he introduced the 'subscription watch' (fig. 347) to make his work available to less wealthy customers. These were watches made in limited series, for which a subscription was paid in advance. Normally almost every watch Breguet made was different from the rest—'subscription watches' were not one-offs, but were constructed, on however rarefied a plane, according to a kind of assembly-line principle.

Breguet therefore deserves to be included in design

histories as one of the first men to take what one might call a truly modern attitude towards the forms of the objects he was responsible for creating. The appearance of the watches he made is not imposed, but is a pragmatic expression of the functions they were expected to perform, and in some instances of the precise circumstances in which they were made. Having analysed how his watches would best serve those who used them, Breguet brought his conclusions together and made watches which were a unitary synthesis. The taut, economical appearance of the watch is a genuine expression of its maker's whole process of thought.

Genuine factory production of watches was pioneered in America during the nineteenth century, and something like assembly-line techniques existed in American watch factories some years before Henry Ford began to make the Model-T. The Hamilton Watch Co., which produced some of the finest American pocket-watches, exemplified some of the slightly contradictory qualities of American watch design at this period. Cases were often gaudy, in the neo-baroque Victorian style (fig. 348), as if to emphasize the value of the mechanism within. But the mechanisms were advanced for their time (the watch illustrated is keyless, an innovation introduced in this form in 1844), and the mechanical quality was extremely refined. Hamilton produced what was probably the finest marine chronometer ever made, and did it entirely by mass-production methods.

The first watch which was truly for the working man was the Swiss Roskopf watch of *c.* 1870. It was a pocket-watch with a nickel case and an enamel dial. This was the forerunner of the inexpensive, mass-market wrist-watch of the present century. A watch like the Smith's Model R.G. 0307 (fig. 349), first put on the market in 1945–6, has a simplicity and sober straightforwardness which till recently would have made it seem the last word in design for the mass market. But the attitudes of the watch industry, and indeed those of their customers, have recently been revolutionized by the arrival of the quartz watch. Its predecessor was the electric watch, which first appeared in practical form as recently as 1957. This

346 (*left*) Timex Quartz digital calculator watch, 1981.

347 (*right*) Subscription watch by Breguet.

348 (*far right*) Hunting cased keyless lever watch by the Hamilton Watch Co.

349 (*extreme right*) Smith's wrist watch, model no. R.G. 0307, designed by R. Lenoir, 1946.

was followed by an electronic watch using a tuning fork to ensure accuracy—the Bulova 'Accutron', which was the first decisive step away from the balance and spring. The tuning-fork in turn was replaced by the quartz crystal which is the heart of the present-day quartz watch. This did two things simultaneously: it simplified the actual mechanism, making it much cheaper and simpler to manufacture, and it provided a degree of accuracy equal or even superior to that of the finest hand-made conventional watches—chronometric accuracy was suddenly available to everyone, instead of being the prerogative of the rich. Especially at first, it was easier to manufacture quartz watches in digital form, with a display provided by light-emitting diodes. This abolished the traditional analogue face, though manufacturers of quartz watches have since brought it back in response to popular demand. It is also possible to equip quartz watches with the complications which were formerly the mark of luxury timepieces, at little extra cost. They have alarm-striking mechanisms, and give the date and the day of the week. It has even been possible to marry an electronic watch to a tiny electronic calculator (fig. 346), or even to a miniature radio.

But these watches also present themselves more and more candidly as throwaway items, certainly not as

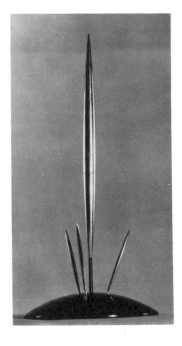

350 (*below left*) The original ball-point pen, patented by Laszlo Biro, 1938.

351 (*left*) Biro in the shape of the 'Vertical Feature' at the South Bank Exhibition, 1951. Designed 1950.

352 (*right*) Standard ballpoint, plus fantasy versions, 1970s.

heirlooms to be passed on to one's children. The cases, for instance, are frequently made of flimsy plastic rather than metal. The attitude cheap digital watches promote is the acceptance of technology as a matter of course.

The plastic watch does not as yet rival, in terms of expendability, another very typical plastic artifact of our time—the ballpoint pen. A primitive form of ballpoint was known as early as 1888, but in the form in which they are now familiar to us ballpoints derive from a patent taken out by Laszlo Biro in 1938. Biro's patent, like the design of the digital watch, consciously rethought a familiar if not particularly ancient everyday object. The fountain-pen, till then firmly established as the most convenient and portable writing implement, was suddenly old-fashioned. In a ballpoint it was not merely that the nib was abolished—it used viscous rather than fully liquid

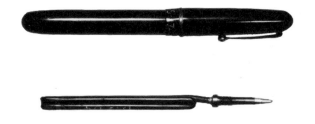

ink, and was fed by means of a spring-loaded plunger which was quite different from the normal fountain-pen mechanism. Rather than refilling a ballpoint, one threw the whole mechanism away once the ink was used up. After the manner of many innovatory designs, ballpoints at first followed established convention—that is, they tried to look like the fountain-pens which were their more expensive rivals (fig. 350). But the working parts were so cheap to manufacture in quantity that the device was soon seized upon by souvenir manufacturers and vendors of cut-price novelties. Ballpoints were transmuted into the strangest forms. It was perhaps fairly logical to manufacture one to celebrate the South Bank Festival of Britain Exhibition of 1951 (fig. 351), especially as it could take the form of the 'vertical feature' more popularly known as the Skylon. More recently, however, ballpoint pens have been disguised as everything under the sun—as corn-on-the-cob, as celery stalks, even buried inside a realistic looking potato (fig. 352). Though not always very practical in themselves, these strange forms proclaim the

ballpoint's place in mass culture — the cheapest writing implement making a permanent mark till the appearance of rivals like the felt-tip pen, easily disposable, almost without personality in the basic form which is also shown here for comparison. Marcel Duchamp said of his 'Ready Mades', themselves industrial objects picked out from the available mass of such things, that they existed more solidly in 'one's grey matter' as works of art than they did when one actually studied them. One might almost say that ballpoints have achieved this ideal transparency— they offer a service rather than existing in their own right, and the cheapest digital watches are well on the way to sharing the same state. Writing with a ballpoint is almost like writing with one's fingertip; wearing a plastic ditigal watch is almost like having the right time tattooed on your wrist.

Another familiar object which has significantly transformed itself in recent years is the camera. It had already reached the first turning point in its development when the cumbersome cameras used by professionals were joined by the simple roll-film box

353 Kodak folding roll-film camera, 1916.

354 (*far right*) Leica prototype, 1913.

355 (*far right*) Hasselblad 500 CM roll-film camera, 1970.

356 Canon F-1 single-lens reflex camera, 1981.

camera intended strictly for the amateur. The next step was the appearance of a somewhat more sophisticated roll-film camera with a bellows and focusing device (fig. 353). The example illustrated dates from *c*. 1916. Another and even more significant step was taken at about the same time—the introduction of the compact precision Leica (fig. 354) using 35mm film of the kind developed for use in ciné cameras. This was a 'serious' camera, equipped with an excellent lens. It made available a kind of freedom and flexibility in image-making which had never been possible previously for either the professional or the amateur. A whole generation of photographers, chief among them Henri Cartier-Bresson, worked in a new style of ruthless candour which the Leica made possible.

The Leica was not the last word in camera design. The camera enthusiast who visits a large photographic dealer today will find a very wide range of cameras available at an equally wide range of prices. The most expensive are the big system cameras which produce a $2\frac{1}{4}$-inch square negative—chief among them is the Swedish-manufactured Hasselblad (fig. 355). It is

described as a 'system' camera because there is a basic camera body to which a very wide range of lenses and accessories can be fitted. Unlike the cameras discussed previously, the Hasselblad allows the photographer to see the image he is going to photograph through the actual lens itself. Yet for all its sophistication and refinement the Hasselblad still conforms to the basic Victorian pattern—it is a box, now made of metal rather than wood—with a lens fitted to the front of it.

The common spectrum also covers single-lens reflex cameras using 35 mm film where the image is viewed through the lens as with the Hasselblad, but usually at eye-level (fig. 356). These, too, are system cameras to which a wide range of different lenses and accessories can be fitted. SLRs, which were first introduced in the late 40s, have become totem objects, to the point where one finds their characteristic forms being facetiously imitated, for example, by a novelty cushion (fig. 362). A more serious compliment was paid to the format by the Sony designers who were asked to find a convincing form for the new and revolutionary Mavica camera, which records an image not on film but on a disc which can afterwards be played back on video (fig. 357)—a perfect example of the way in which new and revolutionary technology gains acceptance by aping a familiar form.

In some ways, however, the most significant recent design thinking in this field has gone into cameras intended almost entirely for amateur rather than professional use. The Kodak Instamatic, introduced in 1962 (fig. 358), was that decade's answer to the box Brownie. It was extremely cheap—at the time of its introduction it sold for just over £5 on the British market, including its carrying case. The body used more plastic than it did metal, and the film came as a cartridge, which could be instantly loaded or unloaded. There was a built-in flash for indoor use, but only two basic shutter speeds.

A more recent model, the compact Olympus XA (fig. 359), occupies a kind of halfway house between a 'silly' camera like the Kodak Instamatic and a good-

358 Kodak Instamatic 100, 1961.

359 Olympus XA 35 mm compact camera, 1979.

quality SLR. It is a viewfinder camera with a fixed wide-angle lens of excellent quality, designed, like the SLRs described above, to take standard 35mm film. But like the Instamatic it has a built-in flash, though one of a far more sophisticated type. The striking features of this camera are its extreme compactness combined with ease of handling. For the traveller, it serves as a kind of visual notebook, kept always conveniently to hand. Its lack of bulk means that it can be slipped into a handbag, or into any fairly capacious pocket.

Smaller still, and directed even more firmly at a leisure and holiday market, is the miniature Minolta Weathermatic (fig. 361), which uses easy-loading cassette film. Its particular feature is that it is designed to be resistant to water and sand, and can therefore be used on the beach. Plastic is almost the only material used, and the slightly rounded forms chosen by the designer express the nature of the material as well as making the camera itself more pocketable.

The most significant camera shown here is probably none of these, but a recent Polaroid (fig. 360). The Polaroid–Land process, invented in 1947, in some

357 Sony Mavica Camera, with lenses and Mavipak discs, 1981.

ways returns the photographer to the days of the daguerrotype, taking photographic technology full circle. The camera produces, not a negative from which a positive must afterwards be made, but a unique print. This print is automatically developed, and becomes visible and available to the photographer only a few seconds after he pushes the button that activates the shutter. Edwin H. Land, who invented the process, wanted, as he said, to make the technical part of picture-taking 'non-existent' for the photographer: 'In short, all that should be necessary to get a good picture is to *take* a good picture, and our task is to make that possible.' His philosophy has been pushed even further since his process was first introduced. The latest Polaroid camera features not only automatic metering, but automatic focus. The flash (which is of course built into the camera body) is coupled to the metering system so as to blend with ambient light and produce precise, accurate colour images which develop in the light within 90 seconds of having been ejected from the camera body. The relatively simple plastic housing which contains the electronic circuitry required does not visually express the nature of what happens—in fact, how could it?—the user takes the whole thing on trust, as a piece of magic, which is clearly what Dr Land himself envisaged from the beginning. It is precisely *because* the technology is too complex to be grasped that the user is set free from it. What matters to him is that he has been given the means of recording what he sees almost instantaneously, with an optical instrument which, for all its magical complexity, is now so cheap as to be within the reach of almost everyone, The criterion for the success of the Polaroid is not its looks, nor even the way in which it conforms or fails to conform to one's previous image of what a camera is, but the way in which it lives up, or fails to live up, to the manufacturer's claim that it provides absolutely trouble-free picture-making. Here too, as with the digital quartz watch, one concentrates on what it does, not on its identity as an object.

The technological toys described in this chapter have been around for a long time, if one dates their emergence from the invention of the pocket-watch.

360 Polaroid Autofocus 660 Land Camera, 1981.

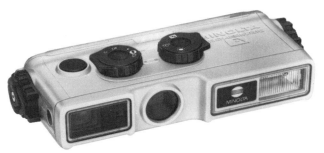

361 Minolta waterproof Weathermatic camera, 1978.

Another way of describing them would be to call them technological companions, in the sense that they are constant reminders of the possibilities technology makes available. An important part of their history is the progression from what is rare and valuable to what is so cheap, as well as so commonplace and accessible, as to be instantly disposable when and if it ceases to function perfectly. Once the designer gets rid of the idea of value, he is able to shift ground, and concentrate on the notion of fluency in use. If the perfect ballpoint pen is an extension of one's finger, then the perfect camera is one which records precisely what one sees – and in the very way in which one sees it—at the push of a button. Designers, therefore, are half-consciously trying to endow the consumer with an additional faculty, rather than with a new possession meant to be added to the heap of possessions which he or she owns already. To identify a thing simply by what it does—that is as part of oneself— rather than in the old way as what it is, through its identity as a separate object in a world of objects, is a very big psychological shift, especially in a society traditionally oriented towards accumulating objects. Design, already transforming itself repeatedly because of the pace of technological change, also has to cope with this aspect of dematerialization, which now seems to be the direction in which technology is going. In fact the question is—does the designer now shape a service, or does he merely continue to give shape to a particular object?

362 Decorative reflex camera in synthetic foam, covered with stretch fabric, designed by Sylvia Libedinsky, 1973.

363 (*opposite*) Snoopyphone, 1981. British Telecom.

Designing to Communicate

It is relevant to what has been said in the last chapter that outlets for communications systems provide the contemporary designer with a great deal of his work. I use this rather clumsy phrase because it is so often forgotten that a telephone, or a radio, or a television set are meaningless objects in themselves, and meaningful only if we think of them in terms of organizational and technological complexities which by their very nature remain invisible to us.

Of the three objects I have just named, the telephone has the longest history. It also bridges the gap between objects which have to accommodate themselves to the shape of the human body, and those where

ergonomic considerations are only secondary. Early telephone designers thought of speaking and listening as two quite separate activities, and designed accordingly (fig. 369). In addition, automatic exchanges were not yet in use, and they did not have to think of ways to accommodate an additional feature, the dial.

A revolution in telephone design took place in the early 30s, and was pioneered in Scandinavia. The Scandinavian countries have small communities which are often not very easily accessible to one another physically, and very early became dependent on the telephone. When automatic exchanges began to

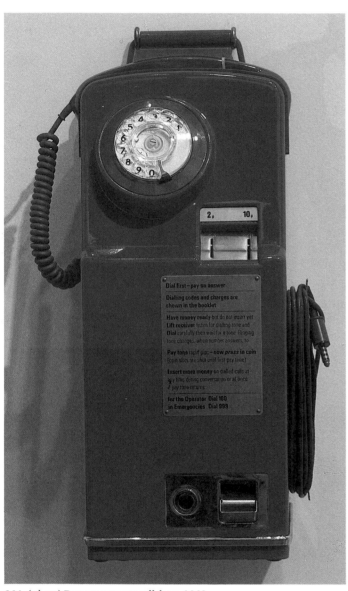

364 (*above*) Pay-on-answer call-box, 1960.

365 (*above right*) Dawn telephone, 1980, British Telecom.

366 (*right*) Cadillac Seville Elegante, 1980.

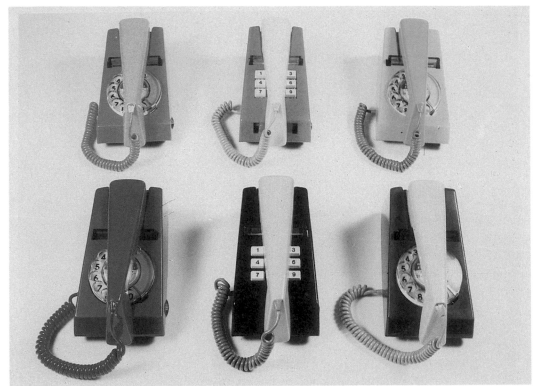

367 (*above*) 'Ambassador' pushbutton telephone, designed by David Carter Associates for G.E.C., 1981.

368 (*left*) Phoenix phones—the Snowdon Collection, 1982. British Telecom.

appear in the early 20s, high development costs meant that it was essential to encourage as many people as possible to become subscribers, and this in turn led to the idea of redesigning the instrument from scratch. The engineers decided to use bakelite, as plastic made it easy to achieve complex curves which were harder to make in metal, but the actual design was the work of a young artist with no engineering background. Jean Heiberg had recently returned from Paris to become Professor at the National Academy of Fine Art in Oslo. The design he came up with had architectural overtones, especially the base, which started out as a classical stylobate, but the total concept was so successful in gaining acceptance from the public that it was exported all over the world, and in Britain various versions of it have continued in current use until the present day (figs. 370, 371). Even where the system of

obtaining a number has been changed, from the dial to the more accurate push-button method, a version of Heiberg's handset is often retained (fig. 367).

However, designers soon became aware of the fact that it might be yet more convenient and economical if all the elements needed were combined in a single form. The first unitary design of this kind was created in 1941, and variations have been appearing ever since, though without ever quite gaining the acceptance given to what still seems the more conventional model (figs. 372, 373). There has also been a somewhat different and more recent tendency. Jean Heiberg designed his instrument to be unobtrusive, instantly acceptable, and to have a shape which explained, without the need for further instruction, how it was to be used. Manufacturers have clearly begun to feel that in this case undemand-

369 Central battery British Post Office pedestal telephone, 1900.

370 (*above left*) Standard British Post Office telephone, 1954.

371 (*left*) Prototype British Post Office telephone, 1954.

372 (*above left*) Ericofon 700 telephone, 1979, manufactured by Thorn Ericsson. British Telecom.

373 (*above*) Eiger telephone, 1981. British Telecom.

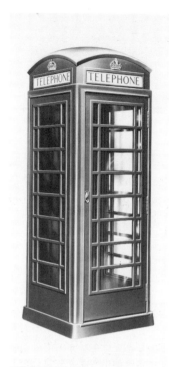

374 Type no. 1 British Post Office telephone kiosk, *c.* 1921.

375 Type no. 6 (Jubilee) telephone kiosk, *c.* 1936.

376 Telephone kiosk designed by Sir John J. Burnet A.R.A., 1924.

377 Swedish telephone kiosk, 1948.

ing acceptance can be taken too far, and have tried to stimulate demand with high-fashion and novelty telephones. British Telecom's Snoopyphone is a rather clumsy tribute to Pop Art (fig. 363). It is interesting to see how in some ways it reverts to the old upright design which the Heiberg model displaced, though still retaining Heiberg's handset. Other new designs have attempted to appeal to the market for glamour, interpreting this idea in rather different ways. The smoothly rounded shapes of 'Dawn' are clearly, as the publicity photograph used here suggests, meant to appeal to a feminine user (fig. 365). One might compare the rounded base to the kind of packaging often used for women's cosmetics. The crisply designed Phoenix phones, by Lord Snowdon, with their angular shapes (fig. 368), were part of the reaction which overtook many kinds of industrial design after a surfeit of bulbous forms in the 50s. These are miniature versions of shapes which can also be seen in a new generation of Detroit automobile designs, for example in the hood and trunk of the Cadillac Seville, Detroit's first

successful luxury 'compact' (fig. 366). But there are practical aspects too—the forms, being simpler, are easier to mould, and the whole instrument is smaller and lighter than its predecessors.

Telephones brought with them a number of subsidiary design problems. The most complex of these were connected with the public, coin-operated phone. There was first of all the need to devise a coin-box mechanism sturdy enough to resist thieves and vandals and simple and reliable in operation. A recent British attempt at this never wholly soluble problem is illustrated (fig. 364). There was also the question of independent housing for public telephones, when these were not to be installed in buildings which already had a major role of their own. In Britain, telephone kiosks evolved from the early 20s towards the 1935 design which until recently remained standard (figs. 374, 375). It is interesting to see how the very term 'kiosk' seems to have exercised a subliminal influence over Sir John J. Burnet, A.R.A., who prepared a design for inspection by the Royal Fine

378 (*left*) Ekco radio, model UAW78, designed by Misha Black, 1938.

379 (*above*) Ekco AD 65 radio designed by Wells Coates, 1934.

380 (*right*) Meridian stereo system, 1981.

Arts Commission in 1924 and provided it with a dome worthy of Brighton Pavilion (fig. 376). By as early as 1948 the practical Swedes had evolved a design (fig. 377) which with its beacon on top and ample space for advertising displays still seems ahead of anything so far proposed by the British GPO.

The radio-set gives the consumer a way of linking himself to a different kind of communication system, where the output is far more varied, but flows in one direction only, as opposed to the two-way communication provided by a telephone. During the pioneering days of radio in the early 20s, listeners used headphones linked to crystal sets. Listening to the radio was a solitary experience, and sets themselves looked like laboratory equipment.

The invention which brought the industrial designer into the picture was the valve-receiver which could be used to power a loudspeaker. This turned listening into a social act—indeed, for a long time people always faced towards the set when they listened, as if it were another person in the room, talking to them. In the late 20s, a radio had come to be regarded as a standard item of home furnishing, and the question of its appearance, as well as its performance, began to be seriously considered by manufacturers. Their first impulse was to make it look like just another piece of furniture. Sometimes, indeed, the set was actually incorporated into a piece which at the same time served another purpose—a grandfather clock or a cocktail cabinet. Cases were made of wood, and the main concession to modernity was the designs used for the loudspeaker grille. A popular one featured a sunrise, in jazz-modern style, symbolic of commitment to the future.

In the 30s the British firm of Ekco began to use distinguished modern architects to design cabinets, and these were commercially successful and improved sales. Serge Chermayeff did a notably simple design in plywood in 1933, and this was followed the next year by Wells Coates's revolutionary design in bakelite (fig. 379). The latter was an early example of plastic

382 Radio by Rondo Gesellschaft, Stuttgart, 1950.

381 (*below left*) HMV Autoradiogram, model 522, 1931.

being used where metal or wood might formerly have been employed, and imposing its own characteristic forms. Its influence can be seen in Misha Black's design of 1938 (fig. 378).

Yet there was always a continuing wish to dress up the radio in some way, to turn it into a decorative object which reflected the social and cultural aspirations of the consumer. An amusing, rather late example of this tendency is the German radio set fitted into a kind of vase (fig. 382), which was evidently intended to look like the modern equivalent of an eighteenth-century *brûle-parfum*.

The real transformation of radio design came about, not through the efforts of eminent industrial designers, but through technological advances which in turn triggered off a fresh wave of changes not only in how radio-sets looked, but in how they were used and in purchaser's attitudes towards them. The invention of the transistor made it possible to miniaturize the set to an extent which the designers of the 30s would have found unimaginable. In August 1955, the Japanese firm of Sony introduced the world's first mass-produced all-transistor radio—the TR-55 (fig. 383). The inno-

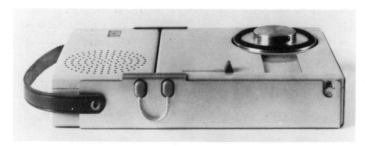

383 (*top*) Sony TR-55 transistor radio, 1955.

384 (*above*) Braun portable radio-record player, designed by Dieter Rams, 1960.

385 (*right*) Sony portable mini stereo cassette player, TPS-L2, 1980.

vation swept the world market. The owner of the set could now carry music and news around with him wherever he went. Manufacturers competed to produce smaller and smaller radios, and technology advanced to keep pace. By 1966, with the Sony ICR-100, integrated circuits had replaced transistors and the radio had shrunk almost to matchbox size.

It began to combine itself with other mechanisms with a similar though not identical function. The German firm of Braun, for example, produced two notable designs which combined a radio and a record-player in a single unit. A battery-operated pocket-size version was designed in 1959—the two parts coupled together for carrying, but could be separated in use (fig. 384). This was the predecessor of the combined cassette player and radio designed to be hooked to the belt and listened to through lightweight headphones which has become an emblem of a free, youthfully independent life-style in the short period since it was first introduced (fig. 385). Another notable Braun design dates from 1962, and also combines a radio and a record-player (fig. 387). Much admired by design professionals for the stringent purity of its outlines, this was also a frank statement of the designer's view that an item of this type remained first and foremost a machine. It was one of the forerunners of the Hi-Tech design fad of the late 70s, and irreverent Braun employees nicknamed it 'Snow White's coffin'.

Yet there is also a sense in which this Braun record-player now seems to fall between two stools. Radiograms had begun their career as assertive pieces of furniture, taking an important place in the living-room (fig. 381). Today, up-to-the-minute hi-fi systems are made up of separate but matched components, designed to be as anonymous as possible (fig. 380). There has thus been a complete reversal of attitudes among purchasers. The early radiograms were prestige items, often in the crudest possible way. New hi-fi systems give the impression that they were meant to be heard but not seen, and this feeling is reinforced not only by deliberate anonymity of design but by a steady shrinkage in cubic volume. One quirk they do still possess, which is that many are endowed with controls which are overcomplicated or actually unnecessary, but which still serve to give the buyer a reassuring feeling of control over a prized but mysterious possession.

During the past 20 years the design of television sets has followed the same general physical and psychological pattern as that of radios. The first all-transistor television set was introduced by Sony in 1959 (fig. 386), only four years after their all-transistor radio, and started the transformation of television from something used for communal viewing, as the radio in the 30s had been a focus for communal listening, into an object of solitary contemplation. Engineers have tackled the problem of miniaturization with much persistence, even though a television set seems in many ways much less suitable for this process of shrinkage than a radio—it can become too small to be seen more easily than a radio can become too small to be heard. At the time of writing the latest solution is the Sinclair Microvision (fig. 5)—a British-made pocket television which combines a 3-inch diagonal black-and-white screen with an FM radio. The compact shape has been made possible by the development of a special flat-screen television tube.

By combining a radio and a television set in the same housing the Sinclair Microvision follows another contemporary trend—that of bringing together two or more functions in the same electronic device. A standard marriage is that of a radio to a digital alarm clock. A particularly elegant recent example is shown (fig. 388), which demonstrates the general tendency

386 The world's first all-transistor television, the Sony TV8–301, 1959.

387 Braun stereo radiogram and radio, 1965.

towards understated unobtrusiveness in all consumer products of this type. Once again the emphasis is now on providing a particular service or group of services rather than an object to be looked at for its own sake.

It might be said that all these electronic devices— transistor radios, hi-fis and television sets—show an increasing preference for unobtrusiveness, almost for anonymity, as if the designer were being employed very largely to ensure that these objects did not catch one's attention in any way at times when they are not in use. When they are in use music, words, even the pictures on the screen become part of the general ambience, an input of expected stimuli—even, one might say, a kind of life-support system whose absence is only noticed by the void left when it is for any reason switched off.

388 Electronic Countdown CD3 radio alarm clock, designed by John Ryan for the House of Carmen, 1977.

Design for Business

The design of office equipment is now quite closely related to the design of the electronic equipment used in the home. In many cases they all belong to the same technological family. The office dictaphone, for example, has undergone the same process of first tidying up, and then miniaturization, as the radio. And one exists to record sound, the other to transmit it. The eponymous Dictaphone Type A, current in 1934 (fig. 390), exposes virtually all its works to the public gaze, including the spare cylinders stored beneath the actual mechanics. When it was redesigned the designer did nothing to the way in which it functioned, but a good deal to improve the way it

looked (fig. 391). It remained, however, a fairly bulky item of furniture.

The Dictaphone 'Time-Master' model, designed in 1947 but not produced until 1955, had shrunk to desk-top size. It was $4\frac{1}{2}$ inches high, $9\frac{1}{2}$ inches wide, and 12 inches long. But it weighed 14 pounds and was therefore not exactly pocketable. It was dependent on mains electricity and offered only 15 minutes of recording time (fig. 392). Other dictating machines of the same generation shared its characteristics, though sometimes they used somewhat different recording systems. The Dictaphone recorded on a 'Dictabelt'; the Agavox of 1955 on a disk (fig. 393).

What really changed the nature of dictating machines was the coming of the battery-powered cassette recorder (fig. 394). The smallest of these were compact enough to be slipped in a pocket, and certainly into a briefcase, and did not require an external microphone. The busy executive could take one with him anywhere. Essentially the process whereby the dictaphone evolved was one in which the designer followed rather than led. He tried to find appropriate forms for the possibilities which technologists made available.

The only thing which impeded the process of miniaturization was the existence of some factor which made it either impossible or inconvenient. Duplicators could not follow the dictaphone in terms of miniaturization, as they still had to cope with standard paper sizes. The first rotary duplicator was introduced by Gestetner in 1903 (fig. 395), and it was manually operated. It has the technical simplicity, directness and functional logic of the best early typewriters. Like them, it kept the working parts exposed so that they were easy to service. The next

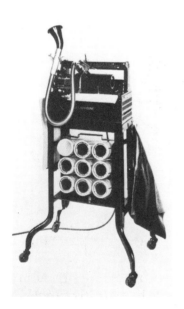

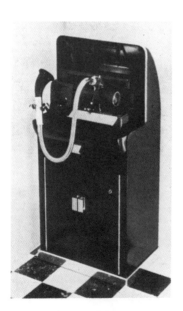

389 (*opposite*) Pelikan desk-top computer by P.A. Design Unit, 1979.

390 (*above far left*) Dictaphone, type A, by the Vitaphone Corp., New York, pre-1934.

391 (*above left*) Dictaphone as redesigned in 1934.

392 (*far left*) Dictaphone 'Time-Master', designed 1947, first produced 1955.

393 (*left*) Agavox dictating machine, 1955.

394 (*above*) Philips 295 Mini Pocket Memo Recorder, 1982.

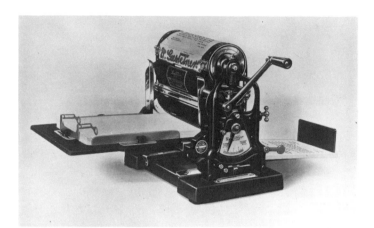

395 Gestetner's first rotary duplicator (no. 3), 1903.

396 Banda Model 90 H duplicator, 1950.

397 Roneo 865 duplicator by H. A. Nieboer, 1966.

398 Gestetner duplicator, 1947.

399 IBM photocopier III, model 60, 1982.

generation of duplicators, which spanned the period running from the 30s to the 50s and even the 60s (figs. 397, 398), essentially represented a tidying-up of the first model. The Banda Model 90H of 1950 boasted of being 'The only British-made spirit duplicator with roller-damping, positive fluid control, flick of the finger master insertion and other exclusive features' (fig. 396). It was still hand-operated, and it continued to be a kind of *ad hoc* printing device, which ran its copies off one at a time from a wax stencil.

These duplicators, though many continue in use, look remarkably primitive when compared to the photocopiers which are now extensively employed (fig. 399). The duplicator with its wax stencil was something whose workings the operator could understand. The photocopying machine remains mysterious, and becomes steadily more so as it becomes ever more sophisticated. Most photocopiers are examples of 'black box' design at its blankest.

An even stranger fate is in the process of overtaking the typewriter. The Underwood No. 1 typewriter of 1897 (fig. 400) was a sturdy basic machine designed to stand up to a lot of hard use. This and similar models set a standard which lasted for half a century, and were subject only to the kind of cleaning-up process which overtook design in the 30s. The first radical change was the electrification of the typewriter (fig. 401). In addition to relieving the typist of physical effort, this brought other improvements as well—for example, mechanical uniformity of touch. But the electric typewriter was very little different from a manual model from the user's point of view. There was another hiatus before the electric typewriter was followed in turn by machines which were not only electric but electronic. These models (fig. 402) did away the conventional array of keys, which was replaced by a golf-ball unit carrying the complete alphabet and any other necessary symbols. More

400 Underwood No. 1 typewriter, 1897. London, Science Museum.

401 Remington electric typewriter, 1950.

402 IBM electronic typewriter, 1981.

recently still, golf-ball typewriters have been challenged by daisy-wheel models which give greater flexibility in alignment and letter-spacing. The daisy-wheel, however, may represent the final stage of the typewriter's evolution, since there is now an entirely new concept—the word processor (fig. 405). Here a use of computer technology enables the operator to record and store a text, and to recall and correct any part of it at will. Word processors are already undergoing the ritual process of miniaturization. Sony recently introduced what they have dubbed a Typecorder (fig. 403). This is a device which will fit without difficulty into a standard briefcase, and which can record up to 120 standard pages of text on a microcassette. Its remaining link with the typewriter of the late Victorian period is the fact that it still makes use of a standard typewriter keyboard.

The Typecorder is closely related to a new generation of 'personal' computers—personal because they are of miniature size. The Sinclair ZX Spectrum measures $8\frac{1}{2}$ inches by 5 inches by 1 inch, and can have as much as 48K RAM memory capacity. It is the big brother of the even smaller Sinclair ZX81, which the manufacturer claims has sold more than 350,000 units world-wide. The ZX 81 measures only 6 by $6\frac{1}{2}$ by $1\frac{1}{2}$ inches and weighs 12 ounces. Its entire circuitry is contained in a single master-chip (fig. 407). Sinclair design is marked by elegant simplicity. Each computer is a black plastic tablet, which presents a set of keys to the user. This format has in fact become a standard feature of the design of small computers—the Pelikan desk-top computer from Germany is very similar to the Sinclair models manufactured in Britain (fig. 389).

Computer technology now enables machines to undertake tasks which would have been considered impossible only a short time ago. Some of their functions are so complex that it still seems astonishing

403 (*opposite above*) Sony Typecorder, 1981.

404 (*opposite below*) Microprocessor-controlled NC-scriber drawing-pen by Rotring, 1980.

405 (*left*) The Nexos 2200 Word processor designed by Richard Satherley, Satherley Associates, 1980.

406 (*below*) Mini-computer for Interset Computer Systems, designed by Peter Harries Associates, 1978.

that they can be carried out mechanically—a case in point is the microprocessor-controlled drawing pen by Rotring—the NC-Scriber. Here, too, the appearances of the machine is hardly expressive of what it can accomplish (fig. 404).

Perhaps it is a reflection of the astonishment felt by the designers themselves that some computer designs carry inexpressiveness to a deliberate extreme (fig. 406). The box with its discreetly ranged set of keys yields its secret only to the thoroughly instructed and initiated. In fact, given the nature of microchips and of computer circuitry in general, it is in any case very difficult for the designer to seek for an expressive form—one which immediately makes it clear not only what kind of mechanism lies within, but what the general nature of its capabilities are and how it is meant to be used. The result is often a certain confusion about the nature of the machine itself. The

407 Sinclair ZX Spectrum mini-computer (right) and the ZX 81.

deliberately inexpensive mini-computer shown might just as easily form part of a modern hi-fi system—it is another extreme example of 'black box' design.

Nevertheless, it must also be recognized that the industrial designer's role in creating such things has in fact altered to a remarkably small extent since Raymond Loewy began his career in the 30s, though the actual technology may now be much more advanced. He is still engaged in putting a practical and attractive envelope round something whose workings he may understand only in a fairly rudimentary way.

To accomplish his task successfully he has to think of two things—ergonomics in the broad sense (that is, not only the way in which human bodies are constructed but about things such as reaction time); and what the object itself is supposed to accomplish. His aim is to harness the user to the used in the smoothest, simplest and most painless way. This means taking into account mental states as well as physical facts. Office machines, like machines in the home or even in the factory, need less and less physical effort on the part of the user. But a machine will be tiring, or annoying, or muddling to use if it is not possible to grasp quickly and easily a basic principle of use. Too many designs for office equipment fail because the equipment is efficient once you have mastered it, but impossible to fathom if you are unfamiliar with the way it operates—which means a tedious hiatus every time a new operative has to be trained. An important part of modern design work is, therefore, to discover ways of seeing to it that the object educates the user in terms of its own use. This in turn means that the designer is often the traditionalist as well as the innovator in a team which yokes the designer on the one side to the technologist or engineer on the other. The engineer is anxious to create *ab initio*; the designer, perhaps surprisingly, must ask himself what is established in this particular field, and how people use it. It is much easier to teach someone to use a new machine if they can make a connection with a machine they already know how to use—one reason perhaps why the modern word processor still religiously preserves the traditional arrangement of letters on a typewriter keyboard, and even, for that matter, the keyboard itself.

Designing the Package

Lettering and packaging occupy a kind of no man's land in design history. Good lettering is essential to many aspects of the designer's job, and it is perhaps the despair of contemporary practitioners that so much of the best signposting is by no means contemporary and is the work of that great designer, Anon (figs. 409 and 410).

Even in the period before the growth of the modern advertising industry, labels and other packaging were important to the shopkeeper. The two early examples illustrated (figs. 411 and 412) represent the birth of the whole packaging industry as we now know it. At first sight it is impossible to discover very much in common between pins and sago powder, but in fact the two commodities face manufacturer and retailer with a similar range of problems. For example, in each case it is difficult to be sure of the source (which is also the guarantee of quality) without an accompanying label. These early suppliers have adopted an identical strategy—the label not only states what the commodity is and who is responsible for supplying it, but an attempt is made to identify something humble and commonplace with the noblest in the land. The sago powder label therefore carries the British royal arms, and the pin-maker proclaims himself to be the supplier of 'Her Royal Highness the Duchess of York'.

The German Werkbund, still the source of so much of our thinking about the nature of design, interested itself in packaging and labelling, and reproduced approved specimens in its annual yearbooks (fig. 413). It is interesting to compare these rather stolid Germanic examples with some of the packaging which was being produced in Britain at about the same time—for instance, the elegantly restrained Art Nouveau favoured by the firm of Yardley for its cosmetics (fig. 414). The contrast is informative in several respects—it shows how, long before the days when the industrial designer achieved full recognition, the people concerned still instinctively thought in terms which would be accepted in leading design studios today. The purpose of packaging is not only to make the product recognizable, and to

408 (*opposite*) Sainsbury's self-service packaging, 1967.

409 (*top*) Street sign in Pudsey, Yorkshire.

410 (*above*) Milestone between Evesham and Moreton-in-the-Marsh, early 19th century.

411 (*right*) Pin-paper, *c.* 1795–1820.

412 (*far right*) Sago powder wrapper, *c.* 1780–90. London, Victoria and Albert Museum.

establish a particular manufacturer or retailer's connection with it, but to generating appropriate feelings about the product in a potential purchaser.

The most instructive thing, where packaging is concerned, is to take a single case history and trace it through several decades. The series of illustrations which follows comes from the archives of a great British supermarket chain, Sainsbury's. It illuminates

413 (*below left*) Labels and packaging for the firm of Günther Wagner, Hanover, 1912.

414 (*left*) Powder pack for Yardley & Co., *c.* 1912.

packaging problems and realities especially clearly because so many of these packages are containers for commodities which are completely anonymous when *not* in a package. Raspberry jam spooned on to a plate is simply raspberry jam. Its specific identity as a particular kind of raspberry jam returns only when one tastes it, and then (it must be said) not always reliably.

The earliest group of designs shown dates from the 20s, and they are a surprising mixture. Both the packages and their labels seem to reach back into the nineteenth century. The packet design for 'Selsa' self-raising flour (Selsa was Sainsbury's own brand) has a scrawny, linear quality reminiscent of the style pioneered by Charles Rennie Mackintosh (fig. 417), while the pie-box lid carries an illustration of naturalistic lushness (fig. 415). Much more clearly reminiscent of the period, from the standpoint of today, are the rather brutal sans-serif capitals on the Selsa chicken-soup tin—a distant echo of Bauhaus principles (fig. 418).

By the end of the decade Sainsbury packaging designers were starting to put more reliance on naturalistic illustration, as something which powerfully reinforced the message of the lettering. The lettering itself is a good deal less restrained (fig. 416).

415 Sainsbury's pie-box lid, pre-1920s.

416 (*below*) Sainsbury's jars for better-quality jam, *c.* 1930.

"SELSA"
FINEST
SELF
RAISING
FLOUR
HOME MILLED
3lb. net.
SAINSBURY,
MFORD HOUSE, LONDON, S.E.I.

Restraint returned, however, in the days of rationing and austerity during and immediately after World War II (fig. 420). The élitism of the well-trained professional of that epoch is detectable in certain touches—the typeface chosen for the macaroni packet, for instance, is the elegant but not always very legible Albertus designed for the Monotype Corporation in 1937 by Berthold Wolpe (fig. 430)—a reminder of the fact that packaging is affected not only by practical needs in shopkeeping terms, but also by

420 Sainsbury's packaging, 1947–8.

418 Selsa tins, 1920s.

419 Sainsbury's self-service egg-packs, 1956.
417 (*opposite*) Selsa (sell Sainsbury) flour bag, 1920s.

the resources available within the printing industry to put the chosen solutions into effect.

In the 50s, Sainsbury's, like other supermarket chains, was affected by the new tendency towards self-service selling. This enabled their shops to cope with a broader range of lines, and at the same time enabled them to reduce numbers of staff. The egg-box, designed to reveal what is inside, is an answer to one of the problems created by this switch (fig. 419). More elaborate packs of the same type, designed to show off consumer durables rather than perishables, were being produced in the United States at the same period, and reflected an even more drastic change in American retailing methods. Some of the most ingenious solutions came from Raymond Loewy Associates (figs. 424 and 425). These packs, and other still familiar items which fall into the same category and which were designed at the same time—among them the

421, 422 Exterior and interior of Penhaligon's shop, Covent Garden, London, 1981.
423 (*below*) Packaging for Penhaligon's, 'Victorian Posy' range, 1979.

much admired packaging created for Jacob's Cream Crackers by R. H. Talmadge (fig. 427)—have now taken on a distinctly 'period' look, though at the time when they were introduced they seemed the essence of clean, simple, elegant modernity. The emphasis on line now seems particularly typical of the epoch—the trellis design on the Jacobs' Cream Cracker packet is recognizably a legacy of the Festival of Britain of 1951, where trellis patterns of all kinds ran riot.

This period look should not, however, blind one to a larger issue, which was that of a change of function. In the new self-service stores, a pack not only contained and protected the product, but took over the job of actual salesmanship. It had to tell the customer the brand-name, and also what the pack

contained and its weight. It often had to give instructions for preparation and storage as well—and all this clearly and simply in a restricted physical area. Packs had to be designed so that they could be stacked together easily, and so that they looked as alluring when seen *en masse* as they did when encountered individually. The pack also had to be designed to cope with customer carelessness, in returning things to the shelves upside down or sideways. In these circumstances, the necessary information still had to remain visible.

Sainsbury's chose Leonard Beaumont to be their consultant designer, and changes made to their packaging during the 50s were the result of an effort to solve these problems simply and logically. The results

424 Carton for cooking utensils designed by Raymond Loewy Associates, 1950.

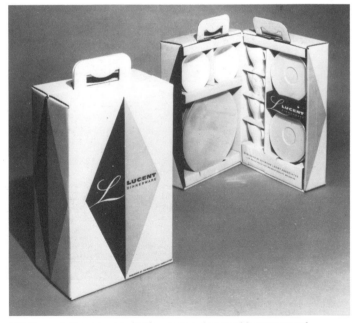

425 Lucent dinner ware display carton designed by Raymond Loewy Associates, 1956.

were practical and worthy, but also sometimes dull.

During the 60s, Sainsbury's again reconsidered their house style, setting up a special 'Sainsbury's Design Studio' at their head office, as part of the existing advertising and publicity department. Designs became bolder. Where an illustration was used, for example on Sainsbury's own-brand corn-flake packet of the late 60s (fig. 426), it was boldly stylized and simplified. Everything possible was done to give the firm's packaging a clean, fresh, healthy look. The vigorous, optimistic graphic idiom, so subtly yet so recognizably different from that favoured in the preceding decade, can be linked to other packaging designs of the time, even when these were intended for completely different products. One can compare the Sainsbury cornflake packet to a

427 First of a range of biscuit packs for Jacobs & Co., designed by R. H. Talmadge, 1956.

426 Sainsbury's cornflakes packet, late 1960s to early 1970s.

428 (*right*) Packaging for IBM Office Products, designed by Paul Rand, *c.* 1960.

429 Sainsbury's own-brand packaging, 1975.

celebrated series of designs devised by Paul Rand for IBM at the beginning of the 60s (fig. 428). With their bold use of the company logo, these are a riposte to the sedateness and over-refinement of the 50s.

During the 70s, Sainsbury packaging designers had to respond to social and economic change, just as they had done in the two preceding decades (fig. 429). For example, a new challenge was the rise in popularity of frozen foods, often purchased in bulk for storage at home. The supermarket gave birth to the much vaster hypermarket. In the huge centres which Sainsbury's opened in conjunction with British Home Stores, goods were colour-coded by department. At the same time, there was something of a revolt against the bold simplicity which seemed the last word in the 60s. Illustrations and photographs were increasingly

important, typefaces were both more rounded and more fanciful. Packaging was losing its importance as a means of conveying basic information, and was taking on the task of giving the consumer a less specific 'good feeling' about the product.

Part of the same tendency is the self-conscious reversion to earlier styles which is a recent phenomenon in certain types of packaging, especially the packaging of avowedly up-market products, sold on a relatively small scale. In extreme cases it is not merely the actual labelling and packaging which are affected, but the whole method of approach to the target audience. A case in point is the small London firm of Penhaligon's, which sells scent, soap, bath-oils etc. The firm's premises in London's Covent Garden were fitted out comparatively recently, but go to great

VERITAS

SCS ALBERTVS MAGNVS
SANCTITATE & DOCTRINA
CELEBER
QVEM PIVS PAPA XI
DOCTOREM VNIVERSALIS
ECCLESIAE DECLARAVIT
IPSE PRO NOBIS ORAT

ABCDEFGHIJKLMNOPQ
RSTUVWXYZ&MT

·ALBERTVS·

THIS FIRST SPECIMEN OF THE ALBERTUS CAPITALS DESIGNED BY BERTHOLD WOLPE IS PRESENTED
TO THE ALBERTVS MAGNVS AKADEMIE COLOGNE
BY THE ENGRAVERS THE MONOTYPE CORPORATION LIMITED AT 43 FETTER LANE IN LONDON
1937

lengths to evoke the atmosphere of the last century (figs. 421, 422). Penhaligon's elegant packaging restates the same theme. Though the firm was genuinely founded in 1870, these designs were commissioned by new and go-ahead owners to evoke the atmosphere which they felt was best suited to the goods they had to offer. As much as the popular television series *Upstairs–Downstairs*, Penhaligon's shop-fitting and packaging are a deliberate but fictional re-creation of a past which never existed in this precise form (fig. 423). Indeed, one of the differences between re-creation and original is that the re-creation is far more consistent in pursuing a particular typographical style.

Typographically, Penhaligon's policy represents a total rejection of an aim which was once universal and unquestioned among leading packaging designers. They all wanted to make a successful marriage between a kind of approved modern taste on the one hand, and commercial effectiveness on the other.

Packaging design becomes doubly significant when one looks at it in the context of modern industrial design taken as a whole. Its ultimate test is not the approval of experts but commercial success. It is so closely intertwined with advertising as to be for all practical purposes indistinguishable from it. Yet it does also invite one to speculate about the nature of design in a number of other fields. It confirms one's impression that the designers of many of the modern electronic items illustrated in this book — clocks, calculators, small computers, radios — are also in fact producing packaging, though of a somewhat more elaborate and ambitious kind than the packs which enclose the items on supermarket shelves. The business of these designers is not with the mechanism, but with the shell, now usually made of injection-moulded plastic. Their contribution to the success of the product is often largely cosmetic. The shell sells the product because its looks are considered pleasing or appropriate, because it suggests (but does not demonstrate) that this object will provide a particular service efficiently, and often because it also suggests that possession of the object in question will confer prestige. A plastic shell for a radio or calculator is thus not in essence very different in function from the wrapping round a packet of digestive biscuits.

430 (*opposite*) Albertus type-specimen, 1937. London, Victoria and Albert Museum.

431 Sainsbury's sultana packet, 1956.

Virtuous Design

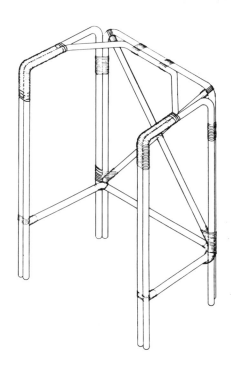

One consequence of industrial design's mixed parentage—on one side the puritan moralism of John Ruskin and his successors, and on the other the hucksterism of modern advertising and salesmanship—has been an increasingly uneasy conscience amongst designers and design theorists. A notable example of this is Victor Papanek's book *Design for the Real World*, first published in 1972. The date of first publication is important, as the early 70s were the moment when the tide of opinion started to turn against immensely successful commercial designers like Raymond Loewy. Papanek's preface is notable for its vehemence. 'There are few professions more harmful than industrial design,' he writes. 'And possibly only one profession is phonier. Advertising design, in persuading people to buy things they don't need, with money they don't have, is probably the phoniest field in existence today. Industrial design, in concocting the tawdry idiocies hawked by advertisers, comes a close second. Never before in history have grown men sat down and seriously designed electric hairbrushes, rhine-stone covered file boxes, and mink carpeting for bathrooms, and then drawn up elaborate plans to make and sell these gadgets to millions of people . . .' These are fighting words, and much of Papanek's book continues in the same strain.

But it is also interesting to see the kinds of things Professor Papanek thinks designers should be producing. The examples he chooses are largely designed by himself and by some of his students. They show a concern with a number of fashionable causes— the disabled, environmental pollution, the peoples of the Third World. A typical design is a non-directional radio meant for use in the kind of Third World country where communications are poor and where there are large numbers of illiterates, as well as lack of power. The radio he devised, with the aid of a graduate student, uses no batteries and no current. The container is a used tin can, which can be decorated according to the owner's taste (figs. 433 and 434). The can contains a wick, fed by paraffin wax or by anything else that will burn, from paper to dried cow-dung. The heat is converted, via a thermocouple, into enough power to operate an ear-plug speaker.

One salient feature of the apparatus is that it is non-directional, and this leads its inventor to a rather startling conclusion. 'In emerging countries,' he says, 'this is of no importance: only one broadcast (carried by relay towers placed about fifty miles apart) is carried. Assuming that one person in each village listens to a ''national news broadcast'' for five minutes daily, the unit can be used for almost a year until the original paraffin wax is used up.' In fact, an inexpensive and extremely effective instrument for thought control, so effective indeed that one wonders why the design hasn't been taken up more widely.

Designers have indeed been increasingly concerned with using their analytical skills in the completely new context supplied by non-industrial countries, and these efforts occasionally produce impressive results. Examples are aids and equipment for the handicapped made to suit local conditions and manufactured from local materials. India has large

432 (*opposite*) Locally made rattan walking-frames, Malaysia.

433 (*above*) Radio receiver for the Third World, made from a used juice can. Designed by Victor Papanek and George Seeger at North Carolina State College. From *Design for the Real World* (Random House/Bantam, 1973).

434 The same radio, decorated with felt and seashells in Bali. From *Design for the real World*.

435 (*top*) The Jaipur limb—climbing a tree-trunk.

436 (*above*) The Jaipur limb—cross-section showing different rubbers.

437 (*right*) The Jaipur limb in use.

numbers of limbless people. The rehabilitation centre at Jaipur Hospital devised a new prosthetic device, the 'Jaipur limb'. This has many advantages over artificial limbs of a more conventional nature. Made of different kinds of rubber (fig. 436), it is rugged enough for the user to go barefoot (fig. 437). It also had lifelike appearance and equally lifelike flexibility—someone equipped with it can climb trees, squat, and sit on the ground cross-legged (fig. 435).

The World Health Organization has issued designs for do-it-yourself crutches made from the simplest local materials (fig. 442), and similar but more elaborate designs for walking frames made of the local rattan have been tried out in Malaysia (fig. 432).

Nearer home there has been considerable research, much of it done in Sweden, into design for the handicapped. A cutlery set in stainless steel has been produced by RFSU Rehab to help people with impaired strength and movement, especially those suffering from rheumatoid arthritis (fig. 438). Also for rheumatic patients is a specially shaped ballpoint pen (fig. 441), which makes an interesting contrast with the fanciful ballpoints illustrated in an earlier chapter. Less specialized are the handle extenders (fig. 439) and the tap-handle extension (fig. 440), intended to overcome a broader range of disabilities.

Projects of this type catch the attention of design

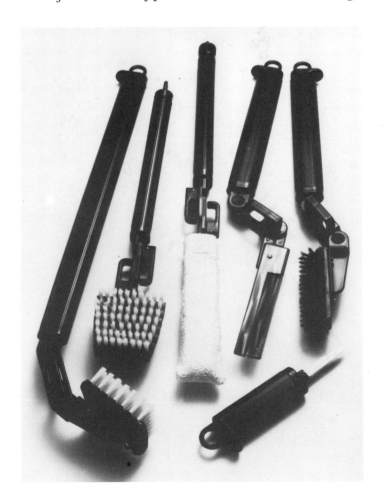

438 Stainless steel cutlery for the disabled designed by Maria Benktzon and Sven-Eric Juhlin, Ergonomi Design Gruppen, manufactured by RFSU Rehab, 1980.

439 (*right*) 'Handy' handle extenders, by Ahlstrom and Ehrich Design for RFSU Rehab.

students with their alluring combination of practic-ality and idealism. Professor Papanek, for instance, has clearly had no difficulty in motivating his students to work with him on the kind of project he himself favours. But if designers dedicate themselves strictly to this kind of task, what happens to the design profession as a whole? The answer must surely be that it becomes, in its present form, an economic absurdity. The busy independent design studios, the design studios within large firms, the design departments affiliated to advertising agencies—all of these are clearly a response to Western consumerism. They feed it, and also feed off it.

Industrial design itself, as this book has attempted to show, existed in embryo long before the Industrial Revolution. The practical problem-solving which is one aspect of design thinking has of course existed in all cultures, including the most primitive. Newer, but still sufficiently venerable, is the restless desire for change which affects all manufactured objects, but particularly those which are personal or domestic. The

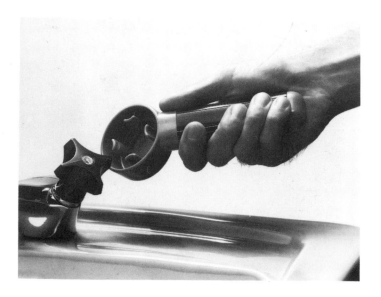

accepted theory of industrial design has always attempted to discount the influence irrational factors have on design thinking. Each succeeding generation, from the Neoclassical period to the present, has tried to defend the favoured solutions of the moment as the only correct and logical ones available. And each generation has been eager to point out the blind irrationality of its predecessors.

In fact, designers have to try to provide answers to problems which are fluid rather than static. They are fluid for two reasons. One is technological change, and this has vastly accelerated since the mid-nineteenth century. The mechanical adding machine, with its levers and its metal and glass casing, is a very different proposition from today's pocket calculator, with its mysterious electronic circuitry encased in plastic. But designers are usually not the prime movers where technological change is concerned. It is not their job to make fundamental inventions, but to discover ways of making these inventions acceptable to the public. That is why so much design is, like Raymond Loewy's first commission for Gestetner, partly something cosmetic, and partly simple tidying up.

The second reason why design problems are fluid is social—society changes just as rapidly as technology. It is difficult, for instance, to decide whether the social or the technological element is uppermost in a modern dishwasher or washer-drier. Technology supplies the means, but the need is a social one—the disappearance of the servant class. Designers are constantly devising new objects to meet needs and deal with situations that at an earlier period did not exist.

Then there are also economic questions. Not merely the question of the financial mechanism which sustains the designer's own career, but the matter of his involvement with the capitalist nexus. Despite the existence of bodies like the Design and Industries

441 Pen for rheumatic patients, designed by Industridesign Konsult for RFSU Rehab, 1979.
442 (*right*) Home-made crutches, from a WHO booklet, 1980.

440 (*opposite*) Tap-handle extension by Ahlstrom and Ehrich Design for RFSU Rehab.

WALK WITH TWO CRUTCHES

HOW TO FIND THE LENGTH OF THE CRUTCHES

- First measure the person to see how long the crutches should be.
- Take a long stick and hold it against the person's side when he/she stands up.
- Place two finger-breadths below the armpit of the person like this and mark the stick at this point.
- The height of the crutches should be the same as the stick from this point to the ground.

HOW TO MAKE THE CRUTCHES

- Now make the crutches like this from strong tree branches.
- Smooth out the roughness of the branches so that they do not damage the skin on the person's hands and side of the chest.

- You can also make crutches like this.

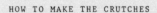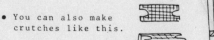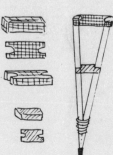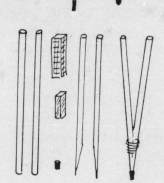

M-153

Association, the design profession did not achieve full recognition until the 30s, and the place where it first received it was the United States—the first society to turn the designer into a kind of culture hero. There can be small doubt that American industry turned to the new race of designers as a last resort, as a possible cure for the Depression, and was to some extent amazed by the results. The high noon of the industrial designer was America from the mid-30s to the mid-50s, and it was at that period that the profession was probably easiest to define. One part of the definition, however, remains deeply unwelcome to those to whom it is applied. It is notable that the leading names of the design generation which flourished in the golden years had backgrounds not in engineering, but in advertising, window dressing and theatre design. One cannot ignore these facts altogether.

Part of the strength of design is not its divorce from fashion, but its alliance with it—and this though it has become commonplace to praise a particular design by saying it 'escapes from fashion'. To escape, it had to be there in the first place. Commitment to change (which is one of the things that fashion at its highest is about) is also important to industrial design.

Faith in technology and its possibilities has not been continuous since the Industrial Revolution first manifested itself. It is not too much to say that one can detect a cyclic pattern—enthusiasm for the machine, even worship of its capacities, followed by suspicion and rejection. In one sense industrial design is as much a product of the negative aspect of this pattern as it is of the positive one. It was born of an impulse to reform and control industry, and was to some extent imposed on the process of manufacture from outside, rather than arising from it. Because industrial objects are made by machines, they are often spoken of as 'impersonal', especially when compared to the products of the hand-craftsman. But they are not impersonal in the reactions they evoke. The history of industrial design is in fact littered with examples of the strong feelings industrial objects arouse.

A major, but sometimes forgotten part of the designer's job is to mediate those feelings—to remain in contact with them, yet to see them for what they are and devise ways of controlling them and channelling them fruitfully. In discussing design, perhaps more than any other subject to do with the history of the visual arts, one is continually faced with the chicken-and-egg situation: did a particular design shape the circumstances in which it was used, or was it purely a product of those circumstances? The really good designer never ceases to put these questions to himself —about his own work and that of other designers. And he does so without giving up the practical analysis of forms and circumstances and practical needs which is also part of his job. But another part of it is indeed to be the professional mediator between the industrial and technological world and that of the consumer, and this, granted the sort of society we have, remains a crucial and influential task.

Bibliography

ASHFORD, F. C., *Designing for Industry*, London, 1955
ASHFORD, F. C., *The Aesthetics of Engineering Design*, London, 1969
BANHAM, REYNER, *Theory and Design in the First machine Age*, London, 1960
BANHAM, REYNER, *Design by Choice*, London, 1981
BAYLEY, STEPHEN, *In Good Shape; Style in Industrial Products, 1900 to 1960*, London, 1979
BAYNES, KEN, *Industrial Design and the Community*, 1967
BAYNES, KEN, AND PUGH, FRANCIS, *The Art of the Engineer*, Guildford, 1981
BEEBE, LUCIUS, *Trains in Transition*, New York/London, 1981
BEL GEDDES, NORMAN, *Horizons*, New York, 1932
BEL GEDDES, NORMAN, *Miracle in the Evening*, New York, 1960
BERESFORD-EVANS, J., *Form in Engineering Design*, Oxford, 1954
BROCHMANN, ODD, *Good or Bad Design?*, London, 1970
CAMPBELL, JOAN, *The German Werkbund*, Princeton, 1978
CAPLAN, RALPH, *Design for America*, New York/Saint Louis/San Francisco, 1969
CARRINGTON, NOEL, *Design in the Home*, London, 1933
CARRINGTON, NOEL, *Design in Civilization*, London, 2nd edn., 1947
CARRINGTON, NOEL, *Industrial Design in Britain*, London, 1976
CHENEY, SHELDON and CHENEY, MARTHA CANDLER, *Art and the Machine*, New York, 1936
CHERMAYEFF, IVAN, and others, *The Design Necessity*, Cambridge, Mass., 1973
DEANE, PHYLLIS, *The First Industrial Revolution*, Cambridge, 1965
DE ZURKO, EDWARD ROBERT, *Origins of Functionalist Theory*, New York, 1957
DOWLING, HENRY G., *A Survey of British Industrial Arts*, Benfleet, Essex, 1935
EDISON INSTITUTE, *Mechanical Arts and the Henry Ford Museum*, Dearborn, Michigan, 1974
FABER, TOBIAS, *Arne Jacobsen*, Stuttgart, 1964
FARR, MICHAEL, *Design in British Industry*, Cambridge, 1955
FEREBEE, ANN, *A History of design from the Victorian Era to the Present*, New York, 1970
FINNISH FOREIGN TRADE ASSOCIATION, *Design in Finland, 1964–1981*, Helsinki, 1981
FRATELLI, ENZO, *Disegno e civiltà della macchina*, Rome, 1969
FREY, GILBERT, *The Modern Chair*, London/New York, 1970
GIEDION, SIGFRIED, *Mechanisation Takes Command*, New York, 1948
GLAESER, LUDWIG, *Ludwig Mies van der Rohe: Furniture and Drawings*, New York, 1977
GLOAG, JOHN, *Georgian Grace*, London, 1956
GLOAG, JOHN, *Victorian Comfort*, London, 1961
GLOAG, JOHN, *Victorian Taste*, London, 1962
GOOD, RICHARD, *Watches*, Poole, Dorset, 1978
GREENOUGH, HORATIO, *Form and function*, ed. Harold A. Small, Berkeley and Los Angeles, 1947
GREENOUGH, HORATIO, *The Travels, Observations and Experience of a Yankee Stonecutter*, ed. Nathalie Wright, Gainsville, Florida, 1958
HALD, ARTHUR, *Swedish Design*, Stockholm, 1958
HARVEY, JOHN, *Medieval Design*, London, 1958
HARVEY, JOHN, *Medieval Craftsmen*, London, 1975
HESKETT, JOHN, *Industrial Design*, London, 1980
HILL, JONATHAN, *The Cat's Whisker*, London, 1978
HIORT, ESBJØRN, *Modern Danish Furniture*, New York, n.d. [1956]
HOLME, GEOFFREY, *Industrial Design and the Future*, London, 1934
HOWARTH, THOMAS, *Charles Rennie Mackintosh and the Modern Movement*, London, 2nd edn., 1977
HUGHES-STANTON, CORIN, *Transport Design*, London, 1967
HUISMAN, DENIS, and PATRIX, GEORGES, *L'Esthétique industrielle*, Paris, 1961

JERVIS, SIMON, *High Victorian Design*, Ottawa, 1974
JONES, CHRISTOPHER, *Design Methods: Seeds of Human Futures*, London, 1970
Jugendstil, catalogue of an exhibition held at the Palais des Beaux-Arts, Brussels, Oct.–Nov. 1977
KEPES, GYORGY (ed.), *The Man-Made Object*, London, 1966
KLINGENDER, FRANCIS D., *Art and the Industrial Revolution*, St Albans, 2nd edn., 1972
KOWENHOVEN, JOHN A., *Made in America*, Newton Centre, Mass., 1957
LETHABY, W. R., *Philip Webb and His Work*, London, 1955
LETHABY, W. R., *Form in Civilization*, London, 1957
LOEWY, RAYMOND, *Industrial Design*, London/Boston, 1979
MADSEN, S.T., *Sources of Art Nouveau*, New York, 1975
MANG, KARL, *History of Modern Furniture*, London, 1979
MCILLHENY, STERLING, *Art as Design; Design as Art*, London/New York, 1970
MEIKLE, JEFFREY L., *Twentieth Century Limited*, Philadelphia, 1979
METZGER, CHARLES R., *Emerson and Greenough: Transcendental Pioneers of an American Aesthetic*, Berkeley and Los Angeles, 1954
MØLLER, SVEND ERIK, *Danish Design*, Copenhagen, 1974
MOSS, ARTHUR, *Successful Industrial Design: Its Creation by Good Management*, London, 1968
Museum of Modern Art, New York, *Introduction to Twentieth Century Design*, New York, 1959
NEUTRA, RICHARD, *Survival Through Design*, New York, 1954
DE NOBLET, JOCELYN, *Design*, Paris, 1974
PAPANEK, VICTOR, *Design for the Real World*, London, 1972
Paris: Musée des Arts Decoratifs, *Formes Industrielles*, catalogue of an exhibition, June–Oct, 1963
PEVSNER, NIKOLAUS, *An Enquiry into Industrial Art in England*, Cambridge, 1937
PEVSNER, NIKOLAUS, *The Sources of Modern Architecture and Design*, London, 1968
POTTER, NORMAN, *What is a Designer? Education and Practice*, London, 1969
PYE, DAVID, *The Nature of Design*, London, 1964
RICHARDS, J.M., *The Functional Tradition in Early Industrial Buildings*, London, 1958
Rochester University, Memorial Art Gallery, *A Scene of Adornment: Decoration in the Victorian House*, catalogue of an exhibition, 7 March–13 April 1975
ROSENTHAL, RUDOLF, and RATZKA, HELENA L., *The Story of Modern Applied Art*, New York, 1948
RAE, JOHN B., *The American Automobile*, Chicago/London, 1965
RAE, JOHN B., *Climb to Greatness: The American Aircraft Industry, 1920–1960*, Cambridge, Mass./London, 1968
READ, SIR HERBERT, *Design and Tradition*, Hemingford Grey, 1962
REDMAYNE, PAUL, *The Changing Shape of Things*, London, new edn., 1960
ROSEN, BEN, *The Corporate Search for Visual Identity*, New York, 1970
Saint Louis, City Art Museum, *Product Environment*, catalogue of an exhibition, 24 April–7 June 1970
SCHAEFER, HERWIN, *The Roots of Modern Design*, London, 1970
SCHEIDIG, WALTHER, *Weimar Crafts of the Bauhaus*, London, 1967
SEMBACH, KLAUS-JÜRGEN, *Into the Thirties*, London, 1972
SHARP, DENNIS (ed.), *The Rationalists*, London, 1978
STURT, GEORGE, *The Wheelwright's Shop*, Cambridge, 1923
TEAGUE, WALTER DORWIN, *Design This Day: The Technique of Order in the Machine Age*, New York, 1940
Transactions of the National Association for the Advancement of Art and its Application to Industry, vol. 1, London, 1890
WINDSOR, ALAN, *Peter Behrens*, London, 1981

Index